SpringerBriefs in Environmental Science

SpringerBriefs in Environmental Science present concise summaries of cutting-edge research and practical applications across a wide spectrum of environmental fields, with fast turnaround time to publication. Featuring compact volumes of 50 to 125 pages, the series covers a range of content from professional to academic. Monographs of new material are considered for the SpringerBriefs in Environmental Science series.

Typical topics might include: a timely report of state-of-the-art analytical techniques, a bridge between new research results, as published in journal articles and a contextual literature review, a snapshot of a hot or emerging topic, an in-depth case study or technical example, a presentation of core concepts that students must understand in order to make independent contributions, best practices or protocols to be followed, a series of short case studies/debates highlighting a specific angle.

SpringerBriefs in Environmental Science allow authors to present their ideas and readers to absorb them with minimal time investment. Both solicited and unsolicited manuscripts are considered for publication.

More information about this series at http://www.springer.com/series/8868

Ellen Wohl

Sustaining River Ecosystems and Water Resources

 Springer

Ellen Wohl
Department of Geosciences
Colorado State University
Fort Collins, CO, USA

ISSN 2191-5547   ISSN 2191-5555   (electronic)
SpringerBriefs in Environmental Science
ISBN 978-3-319-65123-1   ISBN 978-3-319-65124-8   (eBook)
DOI 10.1007/978-3-319-65124-8

Library of Congress Control Number: 2017948730

Printed on acid-free paper

This Springer imprint is published by Springer Nature
The registered company is Springer International Publishing AG
The registered company address is: Gewerbestrasse 11, 6330 Cham, Switzerland

Preface

People have deliberately manipulated rivers for centuries in an attempt to enhance specific river functions, including navigation and water supplies, and to limit hazards such as flooding and bank erosion associated with rivers. A narrowly focused conceptualization of rivers as primarily channels that convey water downstream is one result of this history of river management. Although river channels do convey water, they also transport and store solutes, sediment, and organic matter; create aquatic habitat; interact with the adjacent floodplain and underlying hyporheic zone; and support abundant and diverse biotic communities. River management focused only on water conveyance and hazard mitigation has led to a plethora of problems and to loss of river ecosystem services, prompting scientists, managers, and the public to seek more holistic conceptualizations of rivers as ecosystems.

This book summarizes the state of river science with regard to river ecosystems and argues that management centered on a more holistic perspective of rivers will more effectively sustain river ecosystems and water resources. The book is designed to be accessible to students of river science and to river managers, as well as to scientists from diverse disciplines related to the study of rivers. Following the introductory chapter, succeeding chapters summarize river science, the ways in which human activities directly and indirectly alter river ecosystems, and new management approaches for rivers.

In writing this book, I have benefited from decades of stimulating discussions with research colleagues, graduate students, and natural resources managers in the United States and other countries. I feel fortunate to be able to study rivers and to work with people who are passionate about rivers. I would like to thank Sherestha Saini for the invitation to contribute to the SpringerBrief series, which led to the idea of writing about river ecosystems and river management, and Professor Jacqueline King, Dr. Katherine Skalak, and an anonymous reviewer for detailed and insightful comments that improved this book.

Fort Collins, CO   Ellen Wohl

Contents

Chapter 1
Introduction

The basic objectives of this book are twofold. First, the book provides an overview of rivers as ecosystems that exist within the greater landscape of a watershed or drainage basin. Second, the book explores the management implications of conceptualizing rivers as ecosystems rather than simply as channels to convey water downstream. If rivers are viewed solely as conduits for water, then management can be designed to maximize delivery of specified quantities of water at specified times via intensive engineering and flow regulation. If rivers are viewed as ecosystems, then management must be designed to include other outcomes such as maintenance of physical diversity, connectivity, and biodiversity within rivers. Since the final decade of the twentieth century, river management has increasingly emphasized aspects such as diversity and connectivity. This book reviews scientific understanding of rivers relevant to the shifting emphases in river management and suggests pathways to sustain this management shift.

My intent in writing the book is to summarize recent developments across multiple aspects of river science that are relevant to river management. The book provides an interdisciplinary perspective for students and professionals within individual subsets of river science and an overview accessible to those primarily concerned with river management. I believe that this book is needed because paradigms—theoretical frameworks that shape how we think and act—change slowly, in science, management, and society. A paradigm shift among river scientists has been underway for at least two decades. Ecologists and physical scientists increasingly emphasize the interconnectedness of channels with the floodplain and underlying hyporheic zone, the watershed, and the broader world (e.g., Fausch et al. 2002; Muehlbauer et al. 2014; Harvey and Gooseff 2015; Gurnell et al. 2016). Scientists also increasingly highlight the importance of changes through time—floods, droughts, fires—in maintaining a healthy river ecosystem (Nilsson and Berggren 2000; Tockner et al. 2003). This paradigm shift has been slower to reach those charged with managing rivers and other natural resources, as well as society at large.

© The Author(s) 2018
E. Wohl, *Sustaining River Ecosystems and Water Resources*, SpringerBriefs in Environmental Science, DOI 10.1007/978-3-319-65124-8_1

Following this introductory chapter, the second chapter covers basic aspects of physical and ecological science that underlie contemporary scientific understanding of rivers. This information supports and expands on a primary thesis of the book: rivers are most appropriately considered as ecosystems rather than as simple channels for downstream conveyance. The intent of the second chapter is to support this contention by reviewing how physical characteristics of rivers support biotic communities and river ecosystem functions. The third chapter reviews the diversity of human alterations of rivers throughout history in order to provide context for what has been changed and why river restoration is so widely undertaken. The final chapter discusses different forms of research and management that are being used to restore rivers.

First, however, it is useful to consider how different people perceive rivers and what they mean in using the phrases *river health* and *river ecosystem*.

1.1 Perceptions of Rivers

Perceptions and language matter. Consider the jungle versus the rainforest. Perceptions of the jungle have traditionally been largely negative, emphasizing the darkness under the dense tree canopy and the danger hidden within that darkness. This is reflected in expressions such as 'it's a jungle out there' or 'the urban jungle'. The word rainforest, which is used for many of the same tropical forest environments as the word jungle, has come to have very positive connotations. Rainforest is commonly equated with biodiversity and a threatened, fragile environment. A jungle is something to be overcome and tamed. A rainforest is something to be cherished and protected. I start with this anecdote because human perceptions of natural environments are at the heart of how we attempt to manage those environments and whether we emphasize alteration and subjugation of an environment or protection and restoration. Consequently, the next two sections discuss perceptions of rivers and river health.

Rivers are one of the most important and endangered ecosystems on Earth. An ecosystem is a community of living organisms linked together and to the adjacent environment through fluxes of nutrients and energy. Rivers have sometimes been conceptualized as simple channels for the downstream conveyance of water and sediment, but this conceptualization misses the critical aspects of rivers that sustain life by providing ecosystem services. Ecosystem services include the general categories of provisioning, regulating, supporting, and cultural (MEA 2005), and rivers provide all of these services abundantly. Provisioning refers to products obtained from ecosystems. River provisioning comes from water, food such as subsistence and commercial fisheries, and fertile soil in the floodplain. Regulating describes benefits obtained from the regulation of ecosystem processes. Regulating in river ecosystems comes from reduced flood hazards associated with energy dissipation across floodplains. Supporting ecosystem services are those necessary for the production of all other ecosystem services (De Groot et al. 2002; MEA 2005). Examples of supporting ecosystem services from rivers include primary production by plants in the channel and floodplain and nutrient recycling by microbial organisms and macroinvertebrates inhabiting the channel and floodplain. Cultural ecosystem services describe the non-material benefits that people derive from ecosystems: who among us does not enjoy spending time along a river?

Diverse groups of people perceive rivers differently. This can be illustrated in an academic context by simple generalizations about traditional engineering, geomorphic, and ecologic perceptions of rivers (Fig. 1.1; Table 1.1). A traditional engineering disciplinary perspective is likely to focus on how water flowing down a river channel interacts with the channel boundaries to create a distribution of hydraulic force that can modify the channel in ways that may be undesirable for those seeking to use the channel for navigation or to limit bank erosion or overbank flooding. The ideal river from a traditional engineering perspective is one that remains stable, without substantially incising its banks, eroding its bed, accumulating sediment within the channel, or flooding out of the channel. Engineering emphasizes using mathematical relations to predict the channel form best suited to create this stability. Engineers also design channel modifications such as bank protection or instream structures to promote channel stability. Equations describing water flow in an open channel and the mechanics of sediment transport provide the foundation for engineering understanding of rivers.

A geomorphic disciplinary focus is more likely to start with how interactions across the drainage basin influence water and sediment entering the river network. Geomorphic investigations also emphasize how channel form and process result from interactions among water and sediment within the channel, the erosional resistance of the channel boundary, and the stability of relative base level. Geomorphic investigations of rivers include how these diverse interactions occur across differing scales of time and space. Qualitative and quantitative conceptual models of equilibrium and nonlinear dynamics that describe river process and form provide the foundation for geomorphic understanding of rivers.

An ecological perspective of rivers is most likely to focus on how interactions between abiotic factors and biota create fluxes of matter and energy, as well as structuring biotic communities. As with geomorphic investigations of rivers, ecological investigations examine processes and biotic communities across diverse scales of space and time, from entire ecoregions and drainage basins over hundreds to thousands of years, to channel units such as pools and rivers and nutrient uptake at minutes to hours. Conceptual models of longitudinal and lateral patterns of matter and energy fluxes and communities (Fig. 1.1) provide the foundation for ecological understanding of rivers.

None of these disciplinary perceptions of rivers is more correct than the others. They differ in their relative emphases, but they also share many areas of common interest. Ecologists can argue that their perceptions of rivers are the most comprehensive, because they include and build on understanding derived from engineering hydrology and hydraulics and geosciences temporal and spatial scales, but also

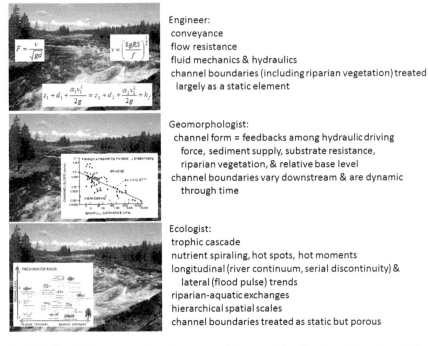

Engineer:
conveyance
flow resistance
fluid mechanics & hydraulics
channel boundaries (including riparian vegetation) treated
largely as a static element

Geomorphologist:
channel form = feedbacks among hydraulic driving
force, sediment supply, substrate resistance,
riparian vegetation, & relative base level
channel boundaries vary downstream & are dynamic
through time

Ecologist:
trophic cascade
nutrient spiraling, hot spots, hot moments
longitudinal (river continuum, serial discontinuity) &
lateral (flood pulse) trends
riparian-aquatic exchanges
hierarchical spatial scales
channel boundaries treated as static but porous

Fig. 1.1 Simplified representation of a river corridor as traditionally viewed from three different disciplinary perspectives: here, the Benbryteforsen (bone-breaker) rapid on the Pite River in Sweden

explicitly incorporate organisms. Engineers and physical scientists also increasingly acknowledge the importance of organisms in rivers, both as drivers of many aspects of river management (e.g., Stewart et al. 2005; Haase et al. 2013) and as ecosystem engineers that strongly influence river process and form (Corenblit et al. 2011; Merritt 2013; Riggsbee et al. 2013). Each disciplinary perspective adds an important piece to a holistic understanding of rivers.

Recognizing these traditional differences in disciplinary perspectives is useful when undertaking transdisciplinary river research or coordinated river management. No traditional disciplinary background effectively covers all aspects of river ecosystems, but each discipline has important insights to offer. Analogously, people living along rivers or using the rivers for their livelihood or for recreation can offer insights that are important to incorporate into river management, not least because these people are the stakeholders who support or do not support river management activities.

Table 1.1 Disparate disciplinary views of a river

Discipline	Structure	Stability	Energy	Form	River	Downstream flux
Ecology						
Aquatic	Organisms	10^0 y	Solar, flow	Hierarchical patches	Channel, hyporheic	Water, organic matter, nutrients, organisms
Riparian	Vegetation[a]	10^0–10^2 y	Solar, flow	Floodplain landforms	Channel, floodplain	Water, propagules, genetic information
Engineering	Dams, bridges	10^0–10^1 y	Flow	Cross sectional, gradient	Channel	Water, (sediment)
Geomorphology	Geologic underpinning	10^1–10^4 y	Flow, tectonics	Bedforms, planform, longitudinal profile	Channel, floodplain	Water, sediment

[a]Canopy height and stem density; ages of plants within a population

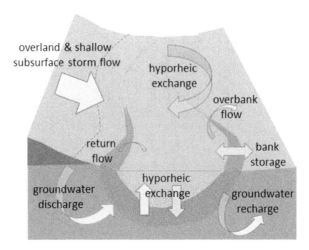

Fig. 1.2 Schematic illustration of the river corridor, including a main channel, floodplain (left boundary outlined by *dashed line*), and hyporheic zone (*orange* fill beneath channel), and highlighting hydrologic exchanges within the river corridor. These hydrologic exchanges represent forms of connectivity in the longitudinal dimension (downstream movement of water), lateral dimension (surface and subsurface movements of water from uplands into the river corridor and between different components of the river corridor), and in the vertical dimension (hyporheic and groundwater exchanges). (Modified from Harvey and Gooseff 2015, Fig. 1)

1.2   Healthy Rivers

Because rivers are ecosystems, throughout this book I refer to the physical environment of a river as the river corridor, rather than just the channel. A river corridor includes the main channel and secondary channels, where these are present; the floodplain; and the hyporheic zone underlying the channel and floodplain (Fig. 1.2). The floodplain is that portion of the valley bottom inundated by water overflowing the channel banks during peak flows that occur every year or every few years. The hyporheic zone underlies the river corridor and is delineated by flow paths originating from and terminating in the channel (Harvey and Gooseff 2015). A river corridor is thus an integrated physical system, which can be distinguished from a river ecosystem that includes the biotic communities within the river corridor. The distinction between river corridor and ecosystem is to some extent arbitrary because physical process and form in rivers are tightly coupled with biota, as discussed throughout this book.

A focus on river corridors emphasizes diverse forms of connectivity within a river ecosystem. Connectivity refers to the transfer of materials (e.g., water, sediment), energy, and organisms between components of an ecosystem. Components in a river ecosystem include the channel, floodplain, and hyporheic zone, as well as different segments of the river network, such as headwaters versus downstream portions. Longitudinal, lateral, and vertical dimensions of connectivity exist among the components of the river ecosystem.

In addition to the obvious downstream movements of water and sediment, longitudinal connectivity includes upstream movements by organisms. Lateral connectivity is present within the river corridor—between the main channel and secondary channels and floodplains, and between the river corridor and adjacent uplands. Vertical connectivity is present within the river corridor over relatively short distances between the channel or floodplain and the hyporheic zone. Vertical connectivity exists over longer distances between the atmosphere and the river corridor and between ground water and the river corridor. Each of these forms of connectivity is explored in greater detail in Chap. 2.

The history of economic development and industrialization in most countries is a history of altering rivers to convert them from ecosystems and spatially heterogeneous corridors to simpler, more spatially uniform channels. People have built levees and drained floodplains, in the process severing connections between the channel and floodplain. Communities and industries have dumped waste products into river corridors, changing water quality, nutrient loads, and the ability of individual organisms and biotic communities to survive within the river. People have regulated flows to reduce flooding and store water for dry periods or to generate hydropower, in the process altering natural river flow regimes. People have channelized rivers to reduce their lateral mobility and used rivers to transport goods ranging from masses of cut logs to immense barges. And, people have relied on rivers for water supplies and fisheries.

The unintended cumulative effects of the long history of river manipulation are increasingly apparent (Williams et al. 2014). Irrigated agriculture has expanded 174% globally since the 1950s and now accounts for ~90% of global freshwater consumption (Scanlon et al. 2007). Widespread nutrient pollution of river corridors that reflects use of agricultural fertilizers reduces river ecosystem processes (Woodward et al. 2012). Indeed, disruption of global nitrogen and phosphorus dynamics has likely exceeded planetary boundaries of a safe operating space for humanity with respect to global environments (Rockström et al. 2009; Steffen et al. 2015). Other widespread contaminants in river corridors include metals, synthetic chemicals such as pesticides and PCBs, untreated human and animal wastes, and pathogens (Meybeck 2003; Wohl 2014). Nearly 80% of the world's population is exposed to high levels of water insecurity resulting from lack of potable water (Vörösmarty et al. 2010).

Decades of intensive river engineering have facilitated greater construction and settlement within river corridors, resulting in substantial increases in flood damages and exposure to flood hazards (Jongman et al. 2012). Reservoirs trap an estimated 26% of global sediment flux to the oceans (Syvitski et al. 2005), so that sediment flux has declined despite accelerated erosion in uplands, causing widespread erosion of river deltas (Syvitski and Kettner 2011). Dams and diversions homogenize river discharge (Poff et al. 2007) and river engineering homogenizes the configuration of river corridors (Peipoch et al. 2015). Both of these forms of homogenization reduce the ability of river corridors to support diverse and abundant biota and to provide ecosystem services (Moyle and Mount 2007). Projected mean future extinction rates of freshwater fauna in North America exceed those of terrestrial fauna by a factor of five (Ricciardi and Rasmussen 1999). Freshwater faunas have become increasingly homogeneous as a result of extirpation of native species, introduction of nonnative species, and habitat alterations that facilitate these two processes (Rahel 2002). Although freshwater fisheries are not as well tracked as marine fisheries, both commercial and recreational fisheries are causing collapse of fish species in many rivers (Pauly et al. 2002; Post et al. 2002; Carpenter et al. 2011).

I have visited river corridors around the world and I repeatedly hear people— river scientists, natural resources managers, concerned citizens—asking what they can do to restore river ecosystems and the services that they provide. Although they may not use these scientific terms, the concerns are universal: how can we make the water clean again, or bring back the fish, or make the river stop eroding its banks without covering the banks in concrete? How can we make the river healthy again?

River health is a nebulous concept. River scientists do not necessarily embrace the phrase and they differ in how to define and assess it (Boulton 1999; Karr 1999; Norris and Thoms 1999). River health continues to be used because it is intuitively appealing and provides a ready analogy for human health. River health can be defined as the degree to which river corridor energy sources, water quality, flow regime, habitat, and biota match the natural conditions (Wohl 2012). This can be very difficult to assess in regions with many centuries of river manipulation by humans, but comparison to natural or reference conditions remains the most widely used standard for any form of ecosystem health (Palmer and Febria 2012). In this context, natural or reference conditions typically refer to the characteristics of the river ecosystem present prior to intensive human manipulation of the environment within the watershed and the river corridor (Wohl and Merritts 2007; Wohl 2011).

The historical timeline of intensive human manipulation varies widely among diverse locations. In many regions of the world, alteration of upland vegetation for agriculture or timber harvest was one of the earliest human activities to alter water and sediment yields entering river corridors. This occurred as early as 8000 years ago in parts of China (Zhuang and Kidder 2014), and is now occurring in countries such as Uganda (Kasangaki et al. 2008). Direct engineering of channels and regulation of river flows typically occurred more recently, but again over varying timespans. At least some of the rivers in Eurasia have been channelized for more than a century (Pišút 2002), whereas river channelization in the United States occurred primarily during the mid to late twentieth century (Goodwin et al. 1997).

Understanding the perceptions that underlie the use of the phrase 'river health' and the processes that create and maintain healthy rivers is critical to management designed to improve the functioning of rivers as ecosystems. The next chapter focuses on the processes that support healthy rivers.

References

Boulton AJ (1999) An overview of river health assessment: philosophies, practice, problems and prognosis. Freshw Biol 41:469–479
Carpenter SR, Stanley EH, Vander Zanden MJ (2011) State of the world's freshwater ecosystems: physical, chemical, and biological changes. Annu Rev Environ Resour 36:75–99

Corenblit D, Baas ACW, Bornette G, Darrozes J, Delmotte S, Francis RA, Gurnell AM, Julien F, Naiman RJ, Steiger J (2011) Feedbacks between geomorphology and biota controlling earth surface processes and landforms: a review of foundation concepts and current understandings. Earth Sci Rev 106:307–331

De Groot R, Wilson M, Boumans R (2002) A typology for the classification, description and valuation of ecosystem functions, goods and services. Ecol Econ 41:393–408. doi:10.1016/s0921-8009(02)00089-7

Fausch KD, Torgersen CE, Baxter CV, Li HW (2002) Landscapes to riverscapes: briding the gap between research and conservation of stream fishes. Bioscience 52:483–498

Goodwin CN, Hawkins CP, Kershner JL (1997) Riparian restoration in the western United States: overview and perspective. Restor Ecol 5:4–14

Gurnell AM, Bertoldi W, Tockner K, Wharton G, Zolezzi G (2016) How large is a river? Conceptualizing river landscape signatures and envelopes in four dimensions. WIREs Water 3(3):313–325. doi:10.1002/wat2.1143

Haase P, Hering D, Jähnig SC, Lorenz AW, Sundermann A (2013) The impact of hydromorphological restoration on river ecological status: a comparison of fish, benthic invertebrates, and macrophytes. Hydrobiologia 704:475–488

Harvey J, Gooseff M (2015) River corridor science: hydrologic exchange and ecological consequences from bedforms to basins. Water Resour Res 51:6893–6922. doi:10.1002/2015WR017617

Jongman P, Ward PJ, Aerts JCJH (2012) Global exposure to river and coastal flooding: long term trends and changes. Glob Environ Chang 22:823–835

Karr JR (1999) Defining and measuring river health. Freshw Biol 41:221–234

Kasangaki A, Chapman LJ, Balirwa J (2008) Land use and the ecology of benthic macroinvertebrate assemblages of high-altitude rainforest streams in Uganda. Freshw Biol 53:681–697

Millennium Ecosystem Assessment (MEA) (2005) Ecosystems and human well-being: synthesis. Island Press, Washington, pp 155

Merritt DM (2013) Reciprocal relations between riparian vegetation, fluvial landforms, and channel processes. In: Wohl E (ed) Treatise on fluvial geomorphology. Elsevier, Amsterdam, pp 220–243

Meybeck M (2003) Global analysis of river systems: from Earth system controls to Anthropocene syndromes. Phil Trans R Soc Lond B358:1935–1955

Moyle PB, Mount JF (2007) Homogenous rivers, homogenous faunas. Proc Natl Acad Sci U S A 104:5711–5712

Muehlbauer JD, Collins SF, Doyle MW, Tockner K (2014) How wide is a stream? Spatial extent of potential 'stream signature' in terrestrial food webs using meta-analysis. Ecology 95:44–55

Nilsson C, Berggren K (2000) Alterations of riparian ecosystems caused by river regulation. Bioscience 50:783–792

Norris RH, Thoms MC (1999) What is river health? Freshw Biol 41:197–209

Palmer MA, Febria CM (2012) The heartbeat of ecosystems. Science 336:1393–1394

Pauly D, Christensen V, Guénette S, Pitcher TJ, Sumaila UR et al (2002) Towards sustainability in world fisheries. Nature 418:689–695

Peipoch M, Brauns M, Hauer RF, Weitere M, Valett HM (2015) Ecological simplification: human influences on riverscape complexity. Bioscience 65:1057–1065

Pišút P (2002) Channel evolution of the pre-channelized Danube River in Bratislava, Slovakia (1712–1886). Earth Surf Process Landf 27:369–390

Poff NL, Olden JD, Merritt DM, Pepin DM (2007) Homogenization of regional river dynamics and global biodiversity implications. Proc Natl Acad Sci U S A 104:5732–5737

Post JR, Sullivan M, Cox S, Lester NP, Walters CJ et al (2002) Canada's recreational fisheries: the invisible collapse? Fisheries 27:6–17

Rahel FJ (2002) Homogenization of freshwater faunas. Annu Rev Ecol Syst 33:291–315

Ricciardi A, Rasmussen JB (1999) Extinction rates of north American freshwater fauna. Conserv Biol 13:1220–1222

Riggsbee JA, Doyle MW, Julian JP, Manners R, Muehlbauer JD, Sholtes J, Small MJ (2013) Influence of aquatic and semi-aquatic organisms on channel forms and processes. In: Wohl

E (ed) Treatise on fluvial geomorphology, Treatise on geomorphology, vol 9. Academic, San Diego, pp 189–202

Rockström J, Steffen W, Noone K, Persson Å, Chapin FS et al (2009) A safe operating space for humanity. Nature 461:472–475

Scanlon BR, Jolly I, Sophocleous M, Zhang L (2007) Global impacts of conversions from natural to agricultural ecosystems on water resources: quantity versus quality. Water Resour Res 43:W03437. doi:10.1029/2006WR005486

Steffen W, Richardson K, Rockström J, Cornell SE, Fetzer I et al (2015) Planetary boundaries: guiding human development on a changing planet. Science 347:736–748. doi:10.1126/science.1259855

Stewart G, Anderson R, Wohl E (2005) Two-dimensional modelling of habitat suitability as a function of discharge on two Colorado rivers. River Res Appl 21:1061–1074

Syvitski JPM, Kettner A (2011) Sediment flux and the anthropocene. Philos Trans R Soc A369:957–975

Syvitski JPM, Vörösmarty CJ, Kettner AJ, Green P (2005) Impacts of humans on the flux of terrestrial sediment to the global coastal ocean. Science 308:376–380

Tockner K, Ward JV, Arscott DB, Edwards PJ, Kollmann J, Gurnell AM, Petts GE, Maiolini B (2003) The Tagliamento River: a model ecosystem of European importance. Aquat Sci 65:239–253

Vörösmarty CJ, McIntyre PB, Gessner MO, Dudgeon D, Prusevich A et al (2010) Global threats to human water security and biodiversity. Nature 467:555–561. doi:10.1038/nature09440

Williams M, Zalasiewicz J, Davies N, Mazzini I, Goiran J-P, Kane S (2014) Humans as the third evolutionary stage of biosphere engineering of rivers. Anthropocene 7:57–63

Wohl E (2011) What should these rivers look like? Historical range of variability and human impacts in the Colorado front range, USA. Earth Surf Process Landf 36:1378–1390

Wohl E (2012) Identifying and mitigating dam-induced declines in river health: three case studies from the western United States. Int J Sediment Res 27:271–287

Wohl E (2014) River pollution. Oxford bibliographies: environmental science. doi: 10.1093/OBO/9780199363445-0003

Wohl E, Merritts DJ (2007) What is a natural river? Geogr Compass 1:871–900

Woodward G, Gessner MO, Giller PS, Gulis V, Hladyz S et al (2012) Continental-scale effects of nutrient pollution on stream ecosystem functioning. Science 336:1438–1440

Zhuang Y, Kidder TR (2014) Archaeology of the anthropocene in the Yellow River region, China, 8000-2000 cal. BP. The Holocene 24:1602–1623

Chapter 2
Rivers as Ecosystems

This chapter reviews the different physical components of the river corridor, the physical processes that create and maintain those components, and the interactions between physical processes and living organisms. Among the critical aspects of physical processes and physical-biotic interactions are transfers of matter and energy and the occasional disturbances that reconfigure the river ecosystem and the processes occurring within the ecosystem. Individual rivers and segments of rivers respond differently to these disturbances and the nature of the response is characterized by how much the river ecosystem changes and how quickly the ecosystem recovers from change.

rivers are dynamic (handwritten annotation)

2.1 Physical Components of the River Corridor

As noted in the first chapter, a river corridor consists of the active channel(s), a floodplain, and the hyporheic zone. Terraces are also present along many river valleys. Terraces consist of former channel or floodplain surfaces now elevated above contemporary flood levels as a result of continued deposition on the floodplain or incision by the active channel (Merritts et al. 1994; Pazzaglia 2013). Because they are above the elevation commonly reached by flood waters, terraces are not considered part of the river corridor in this discussion. Terraces, however, can strongly influence connectivity between the river corridor and adjacent uplands (Baartman et al. 2013).

Erosional and depositional forms created by river processes typically delineate the active channel, which in most regions does not support mature woody vegetation (Osterkamp and Hedman 1977; Arnaud et al. 2015). Exceptions can occur within ephemeral channels in arid regions, which may have widely spaced mature trees within the active channel (Fig. 2.1). The upper elevation limit at which water is contained within a channel distinguishes the active channel from adjacent floodplain areas.

ephemeral. (lasting for a short time) (handwritten annotation)

no trees by rivers usually (handwritten annotation)

© The Author(s) 2018
E. Wohl, *Sustaining River Ecosystems and Water Resources*, SpringerBriefs in
Environmental Science, DOI 10.1007/978-3-319-65124-8_2

Fig. 2.1 Downstream view of mature Acacia trees in the active channel of the ephemeral Kuiseb River, Namibia. Channel is approximately 90 m wide

A river corridor can have more than one active channel where braided or anabranching planforms exist, or where secondary channels are present. Braided and anabranching planforms, generically referred to as multithread channel patterns, consist of multiple, subparallel channels that branch and rejoin downstream (Fig. 2.2). Braided channels typically have minimally vegetated bars between subchannels. Individual sub-channels and bars continually change shape and position during high flows. Mature forest can cover the floodplain between individual anabranches, which commonly have much slower rates of lateral movement within individual sub-channels than do braided channels.

Secondary channels are those that contain flowing water only during higher discharges. These channels are distinct from tributaries, which original elsewhere and terminate in the main channel, and from distributaries on a delta or alluvial fan, which originate in the main channel and then terminate on the delta or fan.

The main channel and any secondary channels present also determine the channel migration zone (Fig. 2.3), which is the width of the valley bottom across which channels can migrate and have migrated under the contemporary flow regime. Although the channel migration zone typically lies within the boundaries of the floodplain, exceptionally large floods can cause the active channel to move beyond the floodplain. Explicitly delineating a channel migration zone is a helpful reminder that natural channels continually experience erosion along some portions of the bank and deposition on point or alternate bars along other portions of the bank, as well as abrupt lateral movements known as avulsions.

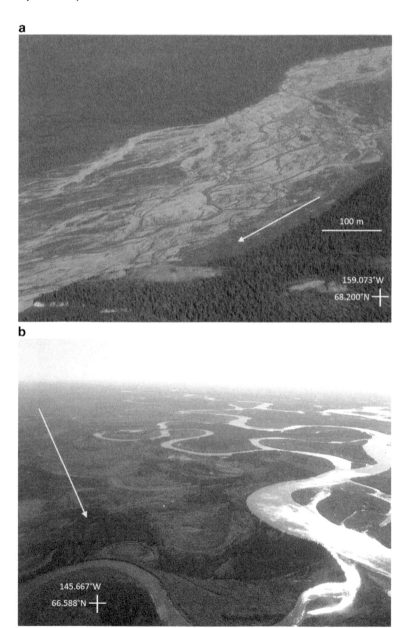

Fig. 2.2 Aerial views of (**a**) Buccaneer Creek, a braided river in the western Brooks Range of Alaska and (**b**) an anabranching portion of the Yukon River in the interior lowlands of Alaska, USA (field of view is approximately 3 km wide and includes only a portion of the active channel migration belt). *Arrows* indicate flow direction

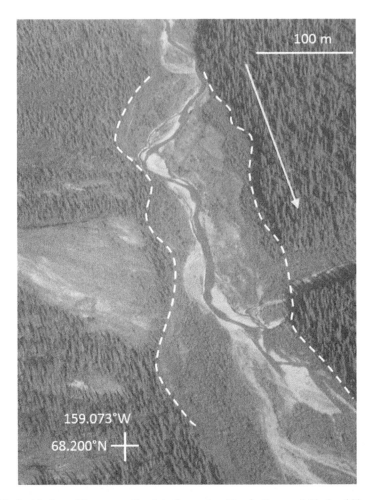

Fig. 2.3 Aerial view of Buccaneer Creek in the western Brooks Range of Alaska. Although the margins of the channel-migration zone are vegetated by shrubs and grasses, the edge of the forest (and the *dashed white lines*) delineate the channel migration zone of the main channel. *Arrow* indicates flow direction

Defining and delineating the active channel are of interest because in countries such as the United States, the active channel of navigable rivers comes under regulatory jurisdiction by the national government, whereas the floodplain does not. The ordinary high water mark defines the boundary between federal jurisdiction and other property ownership. The ordinary high water mark is the line on the shore established by the fluctuations of water and indicated by physical characteristics such as a clear, natural line impressed on the bank; shelving; changes in the character of soil; destruction of terrestrial vegetation; or the presence of plant litter (Wohl et al. 2016). The ordinary high water mark is thus defined based on field indicators, rather than magnitude or return interval of a particular flow. Field delineation of the ordinary high water mark can be very difficult in regions with extremely variable flow, such as drylands or tropical regions, because a recent extraordinary flood may have created erosional and depositional features that obscure the indicators left by ordinary flows (Lichvar and Wakeley 2004).

A floodplain is a relatively flat sedimentary surface adjacent to the active channel that is built by contemporary river processes and inundated frequently (Nanson and Croke 1992; Dunne and Aalto 2013). The lateral extent of a floodplain can be defined in a regulatory context based on the area inundated by a flow with a specific return interval: this leads to definition of a 10-year floodplain, for example, or a 100-year floodplain. River scientists are more likely to describe a floodplain as the surface that is flooded at least once every two years. Rivers with highly variable flow, however, such as those in drylands, can have flat surfaces adjacent to the channel and composed of contemporary river sediments that are flooded only every few decades (Williams 1978a).

In this book, floodplain includes both the floodplain and the riparian zone. The riparian zone is the interface between terrestrial and aquatic ecosystems (Naiman et al. 2005) and, like the floodplain, can be difficult to delineate on the ground. The US Army Corps of Engineers, which oversees regulatory delineations for freshwaters in the United States, defines riparian areas as "lands adjacent to streams …. transitional between terrestrial and aquatic ecosystems, through which surface and subsurface hydrology connects riverine, lacustrine, estuarine, and marine waters with their adjacent wetlands, non-wetland waters, or uplands" (USACE 2012). Riparian zones can include features as diverse as depressions that create floodplain wetlands and higher-elevation natural levees. The lateral extent of the riparian zone can also be defined as the surface extent of frequent flooding and the subsurface extent of mixing of river, hyporheic, and ground waters (Wohl 2014b), creating substantial overlap with the definition of floodplains. Riparian zone is primarily an ecological concept and floodplain is primarily a geomorphic and hydrologic concept. In addition, riparian zones can occur in portions of the river network that do not have floodplains, such as steep sided, narrow canyons.

The floodplain can also include depositional features associated with repeated fluxes of sediment into the main stem river corridor, such as alluvial fans created by tributary streams (Stock 2013) or debris cones created by rockfall or landslides (Harvey 2012). These features can have spatially and temporally complicated interactions with the active channel and floodplain. Fans and cones can expand into the river corridor during periods of large sediment influxes, for example, and then undergo periods of contraction when the active channel erodes the toe of the alluvial fan or debris cone. Where present, fans and cones tend to create at least temporary lateral disconnectivity of water, sediment, and nutrient fluxes into the river corridor, as well as influencing the shape and longitudinal connectivity of the river corridor (Harvey 2012; Fryirs et al. 2007).

The hyporheic zone consists of unconfined, near-stream aquifers where river water is present because flow paths originate and terminate at the channel. This zone can extend several meters below large alluvial rivers and as far as 2 km laterally from the active channel in river corridors with broad floodplains underlain by gravel

(Stanford and Ward 1988). The length and travel time of flow paths within the hyporheic zone vary from a few meters and minutes to hundreds of meters and many hours across the river corridor and downstream, as well as through time in response to fluctuations in river discharge (Gooseff 2010). Hyporheic flow can constitute less than one percent of river discharge in steep, small channels with limited alluvium (Wondzell and Swanson 1996), and 15% or more of surface discharge in larger, lowland alluvial rivers (Laenen and Risley 1997).

Explicitly recognizing different components within a river corridor is inherent in perceiving and managing rivers as ecosystems. If a river is conceptualized simply as a feature for downstream conveyance, then management focuses on the active channel and is more likely to result in severing the active channel from the adjacent floodplain and underlying hyporheic zone. One result of such artificial compartmentalization is that the exchanges of energy, materials, and organisms among the channel, floodplain, and hyporheic zone are likely to be disrupted in a manner that impairs river health.

2.2   Controls on Physical Form and Process in River Corridors

Water and sediment inputs are the primary drivers of physical form and process in all rivers (Fig. 2.4). In forested river corridors, large wood (> 10 cm diameter and 1 m length) historically formed an equally important component. However, centuries of large wood removal in many river networks have reduced the influence of wood and allowed even river scientists to forget how strong an influence wood once exerted on channel and floodplain process and form (Montgomery et al. 2003; Wohl 2014a).

Water flowing downstream converts potential energy to kinetic energy. A river expends flow energy in three ways: overcoming frictional resistance from the channel boundaries and within the flow; transporting sediment; and eroding the channel boundaries. The amount of energy available at any point in space and time reflects the volume of water flowing downstream and the valley and channel geometry, which influence downstream gradient and frictional resistance. The work exerted by the flow reflects the balance between available energy and the erosional resistance of the channel substrate. Substrate erosional resistance results from substrate composition (e.g., sand versus bedrock) and additional resistance created by living aquatic and riparian vegetation and dead vegetation in the form of instream and floodplain wood.

Understanding the manner in which flow energy is used becomes important in the context of river corridor stability. If flow energy increases substantially because of greater discharge, such as during a flood, available energy can exceed the erosional resistance of the channel boundaries, resulting in erosion of the bed or channel widening. Conversely, if flow energy is insufficient to perform the work of

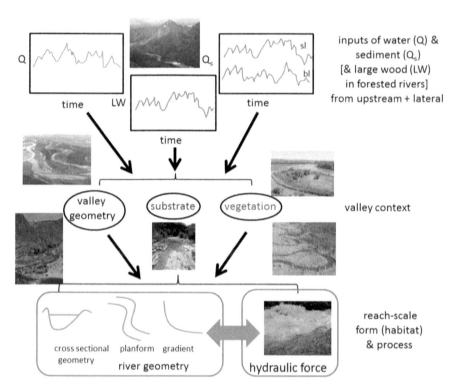

Fig. 2.4 Schematic illustration of the interactions among primary inputs (water, sediment, large wood) to river corridors and the interactions with valley geometry (valley-bottom width relative to channel width; downstream gradient), erosional resistance of the substrate, and vegetation in the river corridor. These interactions result in reach-scale river geometry and hydraulic forces. A reach is a length of river corridor, 10^0–10^3 m in length, depending on the size of the river, with consistent geometry. Each of these variables—water, sediment, and wood inputs, valley context, and reach-scale form and process, fluctuate over varying time scales. In the schematic graph of sediment inputs, sl is suspended load and bl is bed load sediment

transporting sediment supplied to the river corridor, sediment will accumulate along the river, causing bed sedimentation, floodplain aggradation, or channel avulsion.

Sediment entering the river corridor can serve multiple functions. Sediment in transport requires some minimum level of flow energy to remain mobile, but also acts as a tool to erode the channel boundary through abrasion if the sediment is sand sized or larger (Sklar and Dietrich 2004). High concentrations of suspended sediment can increase flow viscosity and dampen turbulence and erosion (Kuhnle 2013), and large amounts of coarse sediment supplied to the river corridor can exceed river transport capacity and form a depositional layer that covers the bed and protects it from erosion (Sklar and Dietrich 2004). Sediment can come from adjacent uplands and enter the river corridor in abrupt, episodic mass movements such as landslides or debris flows (Benda 1990; Korup 2013). Sediment can also move into the river corridor in a more gradual, diffuse fashion with widespread surface runoff or tributary inputs. Finally, sediment can be remobilized within the river corridor from the stream bed, banks, in-channel bars, and floodplain. Detailed sediment budgets indicate that in some rivers, such as the Amazon, more sediment enters the active channel from bank erosion and lateral channel movement across the floodplain than is transported from upstream sources (Dunne et al. 1998). Diverse river drainage basins also receive the majority of upland sediment inputs from a relatively small portion of the total watershed (Meade 2007; Meade and Moody 2010). Understanding the sources of sediment and the processes by which sediment enters and moves within the river corridor becomes particularly important when conditions of sediment deficit or excess create changes in channel form and stability or in nutrient supply.

Processes such as landslides and debris flows can recruit large wood to the river corridor from adjacent uplands (Abbe and Montgomery 2003; May and Gresswell 2003). Large wood can also be recruited from floodplain forests via individual tree fall or mass mortality during blowdowns, insect infestations, or wildfires. Bank erosion also topples floodplain trees, and river erosion exhumes wood buried in floodplain sediment (Benda and Sias 2003). Whatever the source, large wood in the channel increases resistance to flow (Curran and Wohl 2003; Hygelund and Manga 2003) and creates obstructions that enhance hyporheic exchange (Hester and Doyle 2008; Sawyer et al. 2011). Flow deflection around the wood drives local scour of the bed and banks, but also deposition of finer sediment and particulate organic matter in areas of lower velocity associated with the wood (Battin et al. 2008; Beckman and Wohl 2014). Particulate organic matter (>0.45 μm in size) is composed of materials such as fragments of leaves and twigs. Scour and deposition around large wood increase the abundance and diversity of aquatic habitat for a variety of organisms (Richmond and Fausch 1995; Johnson et al. 2003). Large wood can affect the type and dimensions of bedforms (MacFarlane and Wohl 2003) and either increase the spatial heterogeneity of bed substrate by creating patches of differently sized sediment (Buffington and Montgomery 1999), or allow alluvial substrate to accumulate in what would otherwise be a bedrock channel (Massong and Montgomery 2000). Logjams, in particular, can facilitate formation of anabranching channels (O'Connor et al. 2003; Wohl 2011; Collins et al. 2012) and enhance channel-floodplain connectivity and patterns of overbank erosion and deposition (Jeffries et al. 2003).

Although most contemporary studies of wood in rivers focus on small to medium-sized rivers, even the largest rivers historically had substantial quantities of wood (Triska 1984; Montgomery et al. 2003; Wohl 2014a). Large wood continues to influence some large rivers with less alteration of land cover in the drainage basin (e.g., Martín-Vide et al. 2014; Boivin et al. 2015; Kramer et al. 2017).

Floodplain large wood also influences flow resistance and patterns of erosion and deposition during overbank flows, as well as providing vital habitat for diverse types of plants and animals during periods of floodplain inundation and emergence. Case studies illustrating the role of floodplain large wood include the survival of riparian seedlings along the Sabie River in semiarid South Africa (Pettit and Naiman 2006); macroinvertebrate habitat in subtropical coastal plain rivers of

South Carolina, USA (Braccia and Batzer 2001); and small mammals in southeastern Australia (MacNally et al. 2001).

Forested river corridors were much more abundant prior to alteration of land cover, flow regime, and channel-floodplain connectivity. Even ephemeral rivers in deserts of the southwestern United States or central Australia included a forested corridor of trees able to send their roots deep enough to access the water table (Minckley and Rinne 1985; Dunkerley 2014). Wood loads, commonly expressed as volume of wood per unit surface area of active channel or river corridor, were likely orders of magnitude greater in natural forests than in contemporary managed forests and river corridors from which large wood has been continuously removed for centuries (Wohl 2014a). Lower wood loads translate to lower levels of complexity and connectivity within river corridors (Livers and Wohl 2016). Brierley et al. (2005) describe how rapid, systematic clearance of riparian vegetation and removal of instream wood in colonial societies (the Americas, Australia, New Zealand) caused widespread alteration of river form and function in these regions.

A key point in understanding interactions in natural rivers is recognizing that the balance among flow energy, sediment supply, large wood, and channel erosional resistance changes across space and through time. Changes in form and process are inherent in natural river corridors because water, sediment, and wood inputs change across space and through time. These changes result in altered form and function within the river corridor.

The implications of changing water and sediment inputs, in particular, can be explored using three simple examples, one for each of the primary components of the river corridor. The dimensions of the active channel reflect the volumes of water and sediment supplied to the channel. In a channel with an erosionally resistant boundary such as bedrock, only the largest flows will be capable of eroding the channel boundaries. Cross-sectional channel area is likely to reflect the volume of these large flows, even if they only occur once every few decades or centuries (Baker 1988; Wohl 2002). In a more readily eroded channel, the channel cross section may enlarge during high discharges, but deposition of sediment and regrowth of riparian vegetation during subsequent smaller flows may allow the channel cross-sectional area to decrease with time (e.g., Friedman and Lee 2002). The shape and size of the active channel can thus fluctuate over time spans of decades to centuries, but these fluctuations may not result in any net change or trend in active channel geometry over longer time spans. *Channels may fluctuate; but not change over time*

A second example of the implications of changing water and sediment inputs involves floodplains. The characteristics of a floodplain reflect both overbank flows and lateral movements of the active channel through time. Floodplains are built via the combined effects of vertical accretion, or settling of sediment from suspension during overbank flows, and lateral accretion when the active channel migrates laterally across the valley bottom, leaving channel lag and bar deposits in its wake (Nanson and Croke 1992; Dunne and Aalto 2013). Changing water and sediment inputs to the river corridor can accelerate or reduce lateral channel migration rates (Shields et al. 2000; Constantine et al. 2014); increase or decrease overbank flooding and sedimentation (Gomez et al. 1998; Owens and Walling 2002; Miller and

Friedman 2009); and ultimately change the relative importance of vertical versus lateral accretion (Nanson and Croke 1992). The relative dominance of lateral and vertical accretion in turn has implications for river management. Agricultural lands in the floodplain may be enriched by vertical accretion of organic-rich sediment, for example, but largely removed where lateral channel erosion is followed by accretion of coarse-grained channel-bed sediments.

Floodplains along some rivers undergo repeated episodes of aggradation and erosion, as described for high-energy, laterally stable channels in southeastern Australia. Overbank deposition along these channels gradually builds a floodplain of finer sediment during hundreds to thousands of years before a single large flood strips the accumulated sediment to a basal layer of coarser sediment (Nanson 1986). The sequence of floodplain sediment accumulation via vertical accretion then begins again (Nanson 1986).

The final example of the effects of changing water and sediment inputs comes from the hyporheic zone. The location and rate of downwelling from the streambed into the hyporheic zone, and upwelling from the hyporheic zone into the active channel, reflect pressure gradients within the surface and subsurface flow, as well as the porosity and permeability of near-surface bed sediments (Tonina and Buffington 2009; Gooseff 2010). Interactions between river flow stage and ground water pressure dynamics govern hyporheic exchange flows at broad scales. These interactions change with hydrologic conditions and may reverse during dry and wet periods. At the local scale, downwelling occurs at obstacles to flow such as instream wood or beaver dams and bedforms such as riffles and bars, with upwelling downstream from the obstacle or in pools (Buffington and Tonina 2009; Wondzell et al. 2009). Consequently, changes in the configuration of the active channel resulting from altered water and sediment inputs also change the rate and location of hyporheic exchange and this can influence water quality in the channel.

In summary, a healthy river continually adjusts to changing inputs of water, sediment and, in forested rivers, large wood. Figure 2.4 visually summarizes the conceptualization of river forms and physical processes discussed in this book. Understanding the controls on physical form and process is vital for at least two reasons. First, these underlie and support the biotic communities present in river ecosystems. Second, river management can target either the inputs to a river corridor (top row in Fig. 2.4) or the resulting forms and processes within the river corridor (bottom row in Fig. 2.4), but the interactions between inputs and resulting form and process must be recognized if management is to achieve desired outcomes.

2.3 Rivers as Ecosystems

As noted in the introductory chapter, a river is most appropriately conceptualized as an ecosystem because of the close coupling among water and sediment inputs; channel configuration and substrate erosional resistance; biotic communities; water quality; and ecosystem services. Three examples illustrate this coupling: microbial communities in the hyporheic zone; woody riparian vegetation; and beaver.

3 example of interaction

2.3.1 Interconnections Within the River Corridor

Hyporheic exchange exposes nutrients in river water to alternating anoxic and oxic zones in the streambed. These zones are composed of geochemically reactive sediment and microbial communities (Lautz and Siegel 2007). Among nutrients, nitrogen is of particular concern because of human-induced increases in nitrogen entering river corridors and the negative effects of excess nitrogen on water quality and freshwater and marine biotic communities. Only about a fifth of the nitrogen entering rivers is carried to the oceans because of removal and transformation of nitrogen in river corridors (Van Breemen et al. 2002). Much of this removal and transformation depends on biogeochemical hot spots with accelerated chemical reactions, such as are present beneath riparian vegetation (Lowrance et al. 1984) and in the hyporheic zone (Harvey and Fuller 1998). Microbes in the riparian and hyporheic zones are critical to nitrogen removal (Nihlgard et al. 1994), but features such as bedforms and instream wood that promote hyporheic exchange are equally critical to ensuring that downwelling occurs and provides nitrogen to the subsurface microbial communities. In this example, physical features of the river corridor control hyporheic exchanges, but microbial communities control the effects of these exchanges on water quality.

Woody riparian vegetation also illustrate coupling within the river corridor. Woody riparian vegetation strongly influences hydraulics and substrate resistance in many river corridors. The aboveground portion of vegetation growing on the river banks and across the floodplain increases flow resistance. These effects are illustrated by a case study of the sand-bed Rio Puerco channel in New Mexico, USA, where dense woody vegetation along the channel banks reduces perimeter-averaged boundary shear stress by almost 40% and boundary shear stress in the channel center by 20% (Griffin et al. 2005). Vegetation also increases the mass of banks, which can facilitate bank failure. In general, however, riparian vegetation increases the erosional resistance of river banks (Merritt 2013) by creating aboveground frictional resistance to flow that reduces velocity and hydraulic force exerted against the bank, and by increasing the resistance of bank sediment to shearing via the presence of plant roots (Pollen and Simon 2005). High flows that create local bank erosion can remove existing riparian vegetation, but also provide germination sites for new riparian plants. In this example, riparian vegetation both responds to physical processes and alters channel boundaries in ways that influence physical processes.

Beaver (Castor fiber in Eurasia, C. canadensis in North America) provide a third example of coupling within river corridors. Beaver are the premier ecosystem engineers of river corridors in the northern hemisphere. Although now much less common, hundreds of millions of these animals once built dams across rivers and dug narrow canals throughout floodplains, creating river corridor wetlands known as beaver meadows (Naiman et al. 1988a; Polvi and Wohl 2012). Beaver dams create obstructions to flow that facilitate hyporheic exchange and overbank flooding (Westbrook et al. 2006, 2013). Overbank flooding enhances infiltration and maintains a high riparian water table (Gurnell 1998). Water, sediment, and organic matter are stored upstream from dams, in overbank areas, and in the hyporheic zone. This storage attenuates downstream fluxes of water, solutes, sediment, and particulate organic matter (Naiman et al. 1986; Pollock et al. 2007, 2014; Johnston 2014). Water moving across the floodplain and through beaver-dug canals also forms anabranching channels (John and Klein 2004; Polvi and Wohl 2013). By greatly increasing the habitat diversity of the river corridor (Burchsted et al. 2010), beaver increase biodiversity, with beneficial effects on water quality and ecosystem services. Although beaver can only dam small to moderate-sized channels, the animals can construct extensive dam and pond complexes in the floodplains of very large rivers. In floodplains, beaver dam secondary channels, tributary channels, and ground water springs or seeps coming from adjacent hillslopes. In a variety of contexts, beaver influence physical processes and forms in river corridors.

In summary, explicitly recognizing the existence and the details of interconnections within the river corridor is necessary to effectively manage river ecosystems. River management designed to reduce nitrate levels, for example, requires understanding the role of microbial communities within riparian and hyporheic zones and the factors that facilitate exchanges between flow in the active channel and the riparian and hyporheic zones.

2.3.2 Energy Transfers Within River Ecosystems

River ecologists conceptualize coupling between the physical environment and biotic communities in terms of energy flows and disturbance regimes. Energy flows can be illustrated as food chains, food webs, or trophic cascades (Fig. 2.5). Each phrase describes the processes by which solar energy is initially captured through photosynthesis and then transferred among organisms through processes such as herbivory, parasitism, and predation. Within river channels, energy flows among organisms start with either autochthonous or allochthonous production. Autochthonous production occurs within the channel via photosynthesis by bryophytes (mosses and lichens), attached algae and floating or rooted angiosperms (Ward 1992). Allochthonous production is based on organic detritus from outside the channel in the form of dissolved and particulate organic matter. Much of this organic matter entering the channel comes from leaf litter and wood that is broken down by microbes, fungi, and aquatic insects. Within floodplains, energy flows also start with allochthonous production by riparian terrestrial or wetland vegetation, although autochthonous inputs of dissolved and particulate organic matter during overbank floods can be very important.

The primary consumers of organic matter within a channel are biofilm assemblages, macroinvertebrates, and fish. Biofilms composed of attached algae, bacteria,

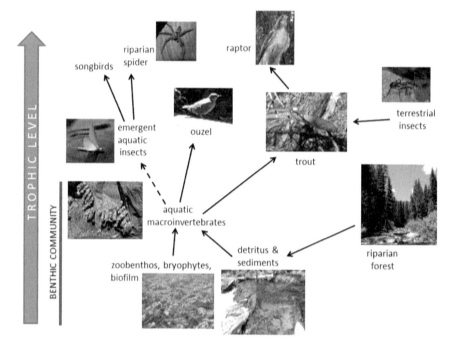

Fig. 2.5 Simplified illustration of the food web in a mountain stream of the Southern Rocky Mountains, USA. Inset images are not to scale and are not necessarily of species present in the Southern Rockies. Benthic community refers to bottom-dwelling organisms in contact with the stream bed

fungi, protozoans, and micrometazoans (Lock et al. 1984) that coat solid surfaces within the channel. These communities of tiny organisms create the slick surfaces that make walking on the cobbles of a small river so challenging. Biofilms are of particular importance because they drive crucial processes within river ecosystems, including cycling of organic matter and primary production (Battin et al. 2016). Environmental heterogeneity of substrate and hydraulics increases the diversity and function of biofilms, providing an example of the interconnections between physical and biogeochemical processes in river ecosystems.

Macroinvertebrates include aquatic insects, crustaceans, mollusks, leeches, amphipods, and nematodes, which collectively typically constitute 98% of the biomass within the channel (Ward 1992). Different macroinvertebrate communities occupy the hyporheic zone, the streambed, the water column, and the air-water interface. Similarly, different species of fish, amphibians, or other aquatic organisms, or different life stages of a single species, occupy different portions of the river corridor. Fish, for example, are commonly migratory on daily and seasonal timespans.

Primary consumers of organic matter in the floodplain include macroinvertebrates and fish that live within floodplain water bodies or migrate onto the floodplain during overbank flooding. Primary consumers also include species that live within the river corridor (e.g., beaver, ouzels, kingfishers) and those that move in and out of the river corridor (e.g., moose or migratory songbirds).

Although distinctive biotic communities occupy hyporheic, channel, floodplain, and upland environments, the boundaries between these environments are leaky, with continual exchanges of matter and energy. Ecologists describe these exchanges as subsidies between aquatic and terrestrial environments. Examples include organic litter entering a channel or floodplain lake from the surrounding riparian forest (Cuffney 1988); aquatic insects emerging from a water body, which are then preyed upon by riparian spiders or birds (Baxter et al. 2005); terrestrial insects such as wasps or ants that are eaten by fish (Nakano and Murakami 2001); deer that move into riparian zones to graze; and salmon that migrate upstream to spawn and die, providing substantial nitrogen inputs to the river corridor via their decaying bodies (Gende et al. 2002). Research continues to discover new dimensions of the complicated energy transfers within river ecosystems and between river ecosystems and the broader environment (Muehlbauer et al. 2014). Clearly, river biotic communities do not and cannot exist in isolation. Changes in energy transfers will result in changes to the abundance and diversity of aquatic and riparian communities, as well as river ecosystem services.

The need to sustain or restore river organisms, including endangered species, drives a significant portion of river management. Effectively managing river biota requires understanding how energy is transferred within river ecosystems and how physical process and form affect energy transfers. Although the basic food webs of rivers have been recognized for many decades, only within the past 20–30 years have ecologists quantified energy transfers between rivers and terrestrial or marine environments. Explicit recognition of these transfers is important to successful river management because river ecosystems do not exist in isolation from areas physically outside of the river corridor.

2.3.3 Disturbance Regimes

A healthy river is not static. Inputs of water and sediment change through time and the river ecosystem responds to these changing inputs. Disturbance regimes characterize these changes in inputs and response in a manner that can inform river management.

Ecologists define a disturbance as any relatively discrete event in time that disrupts ecosystem, community, or population structure and changes resources, substrate availability, or the physical environment (White and Pickett 1985). A flood is the preeminent example of a disturbance within a river ecosystem, although other examples include drought; wildfire in the riparian corridor or adjacent uplands that alters water and sediment inputs to the river corridor; insect infestations or blowdowns in the riparian forest; or debris flows that enter the river corridor from uplands. The disturbance regime is the spatial pattern and statistical distribution of disturbances with respect to magnitude, frequency, and duration of associated changes in the physical environment (Montgomery 1999). River corridors dominated by snowmelt runoff, for example, have a disturbance regime characterized by a single, relatively long duration seasonal flood. River corridors dominated by rainfall runoff, even if the rainfall is strongly seasonal, are likely to have more rapidly fluctuating peak flows.

Disturbance regime in river corridors reflects sources of water, sediment, nutrients, and other materials that originate outside of the river corridor, as well as disturbances within the river corridor. The characteristics of a disturbance and the response of the river corridor to the disturbance also vary within the river network. Headwater channels, for example, commonly lack extensive floodplains and are closely coupled to adjacent uplands. Increased runoff that causes flooding may thus be more likely to change channel form in a headwater channel segment than in a downstream channel segment. All of the excess water during a flood remains within the active channel of a headwater stream, thus creating substantial increases in erosive energy. In contrast, flood waters in a downstream channel segment can spread across the floodplain and dissipate energy in overcoming frictional resistance.

The effect of a disturbance also depends on valley and channel geometry and substrate erosional resistance. Two valley segments in the middle portion of a drainage network can respond very differently to a flood, for example, if one segment has a relatively steep, narrow geometry that concentrates flow energy, whereas the other segment has a lower gradient, wider valley that dissipates flow energy among secondary channels and across a floodplain.

Consideration of disturbance regimes is vital to effective river management because even the most highly engineered and altered river corridor experiences some disturbances. Indeed, physical disturbances such as an annual flood commonly maintain habitat and connectivity that are critical to the survival of plants and animals in the river ecosystem (Junk et al. 1989; Bayley 1991). Traditional river management has sought to minimize natural disturbances and to constrain a river's ability to adjust its form in response to disturbance. This has negatively affected river biotic communities to the point that river management in many cases now seeks to restore disturbances. The three key aspects of disturbance regime for river management involve understanding ① the type, magnitude, and frequency of disturbance within a particular segment of the river corridor, ② how disturbances create and maintain specific features such as channel geometry or channel-floodplain exchanges within the river corridor, and ③ how disturbances influence biotic communities in river ecosystems. Where human alterations have minimized disturbances or constrained the ability of the river ecosystem to adjust following disturbance, river restoration can be targeted at restoring a more natural disturbance regime as in the case of experimental flood releases from dams, or providing space for a river to adjust, as in setting back levees to allow some channel movement during floods.

2.3.4   *Resistance and Resilience*

Resistance and resilience describe the role of disturbances in creating and maintaining physical and biological aspects of river ecosystems. These concepts are useful in the context of river management because they describe how a river ecosystem responds to disturbance and the length of time that changes associated with disturbances are likely to persist.

Interactions between disturbances and river corridor form and erosional resistance govern the diversity and stability of habitats available to aquatic and riparian organisms. In general, greater habitat diversity equates to greater biodiversity in natural rivers (Naiman et al. 1988b), although other factors also strongly influence biodiversity (Ward et al. 1999; Palmer et al. 2010). Resistance can refer to the ability of channel and floodplain substrate to physically resist erosion, but ecologists use resistance in a different context. In an ecological context, resistance describes the ability of a river ecosystem to resist changes in form and process caused by disturbance (Webster et al. 1975). Resilience is the persistence of the river ecosystem and its ability to absorb external changes and maintain the same relationships between biological populations or physical parameters (Holling 1973). Physical scientists also use resilience or sensitivity to describe the ability of a river corridor to return to its pre-disturbance configuration following disturbance (Brunsden and Thornes 1979; Bull 1991; Wohl 2010; Reid and Brierley 2015; Fryirs 2016). In an ecological context, a resistant river corridor undergoes little change during disturbance and a resilient river corridor returns to its pre-disturbance configuration and relationships over a shorter time than the return interval of the disturbance. In other words, a resistant river segment might exhibit very little change following a large flood. A resilient river segment might be changed by the flood, but recover within five to ten years, whereas a flood of that size only occurs once every 50 years.

An important aspect of understanding resistance and resilience is to explicitly recognize that physical characteristics of the river corridor and biotic communities are adjusted to the natural range of variability in water and sediment fluxes through time. Numerous physical characteristics—channel cross-sectional area, the type and dimensions of bedforms, the grain-size distribution of channel substrate, floodplain surface topography and stratigraphy, riparian water table, and so forth—reflect the history of fluctuating volumes of water and sediment entering the river corridor from upstream portions of the river network and from adjacent uplands. Similarly, the species composition, age distribution of individuals within particular species, and abundance and spatial distribution of organisms reflect the history of water and sediment fluxes as well as interactions among organisms. Change in any of these control variables—water and sediment flux, biotic community structure (e.g., via introduced species)—creates the potential for changes in the river ecosystem that can put native species at risk.

Most riverine species are tough: they live in a naturally variable environment and they have evolved numerous mechanisms to survive disturbances such as floods and droughts. However, riverine species are vulnerable to sustained changes associated with multiple stressors, such as competition from introduced species, degraded water quality, and loss of habitat, all of which occur together in many contemporary river corridors. Restoring and maintaining a natural disturbance regime and a river corridor with natural levels of connectivity is the most effective way to enhance the resistance and resilience of a river ecosystem to human-induced alterations.

2.4 Conceptual Models

The influences on and characteristics of river corridors and river ecosystems discussed in this chapter have been conceptualized in qualitative models of rivers. This section discusses conceptual models of river process and form, which have primarily been developed by physical scientists, and conceptual models of river ecological processes and communities that have mostly been proposed by ecologists. There is substantial overlap between these two categories of conceptual models, as discussed in section 2.4.3. Incorporating these conceptual models into river management can facilitate explicit recognition of and emphasis on key factors such as longitudinal or lateral connectivity.

2.4.1 Conceptual Models of Physical Process and Form

One group of geomorphic conceptual models of rivers focuses on how river corridors change through time, particularly in response to disturbances. These models include feedbacks, lag times, thresholds, complex response, characteristic form time, and river metamorphosis, each of which is explained in subsequent paragraphs. The assumption of equilibrium underlies most of these models. In the absence of major external perturbations, the characteristics of the river corridor (e.g., channel width/depth ratio, floodplain surface area, channel sinuosity) exhibit relatively consistent mean values, although there are likely to be continual fluctuations about the mean. This is referred to as steady-state equilibrium (Fig. 2.6).

Equilibrium implies that a change to inputs or controlling parameters will result in a proportional change in the river corridor. If discharge doubles, for example, channel dimensions should increase in a manner that allows conveyance of this larger volume of water downstream. The details of river response can still be difficult to predict precisely: how much of the increased discharge will be accommodated by a larger channel cross-sectional area versus changes in hydraulic resistance?

Feedbacks among components in a river corridor can also complicate prediction of river response to external changes. Positive or self-enhancing feedback occurs when an initial change creates a cascade of subsequent changes that amplify the initial change. Overgrazing reduces riparian vegetation, for example, leading to decreased hydraulic resistance and root reinforcement of stream banks. This allows

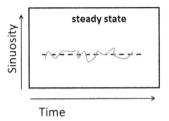

Time

Fig. 2.6 In this schematic illustration of steady-state equilibrium, the sinuosity of the channel varies as individual meanders grow and are cut off (*blue line*), but the average sinuosity (*dashed black line*) does not vary over the time period represented in the plot

high flows to erode the banks, widening the channel and resulting in deposition on bars within the widened channel of some of the newly eroded sediment. The bars deflect current toward the stream banks, resulting in additional bank erosion and bar formation (Trimble and Mendel 1995; Erskine et al. 2012).

Negative or self-arresting feedback occurs when the response of the river corridor dampens an initial change. A tributary fan that laterally impinges on the main channel constricts flow during high discharges. This causes peak flows to transition to critical or supercritical conditions, resulting in high erosive forces that erode the toe of the fan, widening the channel until peak flow returns to subcritical conditions and the channel again becomes stable (Kieffer 1989).

Some river ecosystems exhibit first positive and then negative feedback, as illustrated by ephemeral streams with shallow anabranches in the semiarid tropical portion of northern Australia (Larsen et al. 2016). Wet monsoon forests are present along some alluvial valleys and springs within a landscape dominated by eucalypt savanna. Periods of enhanced stream flow can cause headcuts to form and migrate upstream and this, combined with a highly transmissive shallow aquifer, causes the instream water level and riparian water table to drop. Declining water levels cause the anabranches and formerly saturated, peaty floodplain soil to desiccate. High frequency, low intensity wildfires, combined with soil drying, remove monsoon forest vegetation, which remains intact upstream from headcuts. Increasing stream flows thus create a positive feedback that alters riparian vegetation. However, die-off of monsoon forest trees recruits large wood to the channel. The wood accumulates in logjams that create a local base level, allowing a wet floodplain to re-form and promoting return of wet monsoon forest vegetation in an example of negative feedback (Larsen et al. 2016).

Equilibrium does not imply anything about the rate at which change will occur. A small increase in sediment inputs may result in negligible initial response within the river corridor, but sustained increases in sediment may progressively fill available sediment storage areas in the channel bed, channel bars, levees, and overbank areas, until the channel finally undergoes a substantial change. In this example, a lag time occurred between the initial changes in controlling variables and the response of the river corridor.

Many, if not most, changes in river corridors seem to occur in an abrupt, nonlinear manner described via thresholds. A threshold separates distinct forms or modes of operation of a system. A meandering channel that experiences substantial bank erosion and widening during a large flood can cross a threshold and become a braided channel after the flood (Friedman and Lee 2002). External thresholds refer to change forced by an external factor such as a flood, whereas internal or intrinsic thresholds describe abrupt changes in a river corridor in the absence of external change (Schumm 1979). Relatively small ephemeral channels, for example, undergo episodes of headcut formation and incision that do not necessarily correlate with external changes in water and sediment inputs (Schumm and Hadley 1957). The internal thresholds that trigger headcut formation appear to be caused by deposition of sediment during brief periods of rainfall and stream flow. Deposition in the channel during the falling limb of flow creates a local steepening of streambed gradient that can initiate a headcut during the next period of flow. Once the headcut forms, it can move upstream and trigger a period of channel incision. Incision in the upstream portion of a channel can result in excess sediment downstream that causes aggradation, leading to asynchronous channel changes throughout a single channel or a river network, a phenomenon known as complex response (Schumm 1973; Trimble 2013).

When change does occur within morphologic features such as channel bars or floodplain wetlands, the affected morphologic feature can be described in relation to characteristic form time (Brunsden and Thornes 1979; Bull 1991). A transient form is created by a disturbance, but then modified to pre-disturbance conditions before the recurrence of a disturbance of similar magnitude. A flood causes a meandering channel segment to become braided, for example. The braided segment is transient if it returns to a meandering planform before the recurrence of a flood of similar magnitude. The braided segment is persistent if it remains braided past the recurrence interval of the flood. High-magnitude disturbances are more likely to be primary shapers of river corridors with erosionally resistant channel boundaries. A very large flood creates bedrock channel morphology and boulder bars, for example, which subsequent smaller flows do not have sufficient hydraulic force to modify (Wohl 1992; Cenderelli and Wohl 2003).

River metamorphosis occurs as an abrupt, sustained change in river form (Schumm 1969; Wohl 2013). The original example came from western tributaries of the Mississippi River in the United States. These tributaries crossed the semiarid steppe of the Great Plains as braided rivers characterized by a late spring snowmelt peak flow and very low base flow, with minimal woody riparian vegetation. European agricultural settlement of the region modified flow regime to store the snowmelt peak for gradual release during the growing season for crops. The resulting less variable flow regime allowed riparian vegetation to encroach along the channel banks and floodplain. The Great Plains rivers metamorphosed to meandering or anabranching rivers with dense riparian forests (Williams 1978b; Nadler and Schumm 1981).

A second group of geomorphic conceptual models of river corridors emphasizes patterns through space. Downstream hydraulic geometry, for example, posits that discharge is the dominant control on channel geometry and, therefore, channel parameters such as width, depth, and velocity change progressively downstream as discharge increases (Leopold and Maddock 1953). Geomorphic process domains (Montgomery 1999) and river styles (Brierley and Fryirs 2005) also emphasize spatial patterns in river networks.

As emphasis in river management has shifted to restore more natural river corridors, physical models have emphasized the importance of the natural flow and sediment regimes as primary drivers of river process and form. The natural or altered water and sediment inputs to a river corridor interact with the geometry and substrate resistance of the corridor to determine the types and stability of channel forms present (Fig. 2.4).

The natural flow regime refers to the characteristics of the hydrograph present prior to intensive human alteration of a watershed (Poff et al. 1997) and is commonly described in terms of magnitude, frequency, duration, timing, and rate of change of flow. Numerous methods exist for quantifying how much an altered watershed deviates from the natural flow regime (Richter et al. 1996; Poff et al. 2010). Most of these methods compare discharge characteristics of altered rivers to those of natural rivers. Allowing water to flow downstream in quantities and at times of the year that mimic a natural flow regime can be difficult and expensive, but is increasingly being used in river management (Poff and Matthews 2013).

The natural sediment regime refers to the characteristics of sediment inputs, transport, and storage present prior to intensive human alteration of a watershed (Wohl et al. 2015). Metrics for quantifying deviation from the natural sediment regime do not yet exist, partly because of the extremely limited direct measurements of sediment transport and partly because of the greater difficulty in characterizing sediment inputs and transport relative to river discharge. Wohl et al. (2015) suggest focusing on a balanced sediment regime that involves managing for a desired balance between sediment supply and transport capacity in order to maintain specific physical processes and forms within the river corridor. Wilcock et al. (1996) develop a framework for evaluating sediment supply and transport capacity in relation to maintaining desired channel characteristics. Schmidt and Wilcock (2008) propose numerical metrics for assessing the presence and magnitude of sediment surplus or deficit.

In summary, conceptual models of physical process and form in river ecosystems provide a framework within which to characterize the physical integrity of rivers. This can be useful in understanding how a particular river network or river segment functions through time and in designing management to restore and maintain river health.

2.4.2 Conceptual Models of Ecological Process and Form

As with physical conceptual models, incorporation of ecological models can enhance river management by explicitly recognizing the spatial and temporal patterns within river ecosystems and the processes that drive these patterns. Management incorporating the flood-pulse model, for example, recognizes that periodic inundation of the floodplain is vital to maintaining fish and other river organisms. This section briefly reviews some of the major conceptual models of river ecosystems.

The most widely used ecological conceptual models of rivers emphasize one of three characteristics: longitudinal patterns of process and form; lateral connectivity within river corridors; and changes through time. The river continuum concept (RCC) first described longitudinal patterns (Vannote et al. 1980). The RCC posits progressive downstream changes in the relative importance of primary production versus respiration and in the structure and function of aquatic communities (Fig. 2.7). Ward and Stanford (1983, 1995) describe a variation in these downstream trends associated with the presence of dams and reservoirs that reset the downstream patterns, as described in the serial discontinuity model. Other studies emphasize hierarchical patch dynamics, which characterizes river corridors as consisting of relatively homogeneous patches from the scale of microhabitat up to channel reaches, with distinct changes in process and form between patches (Pringle et al. 1988; Poole 2002; Thorp et al. 2006).

Longitudinal patterns of process can also be described in the context of nutrient spiraling. Nutrients such as nitrogen and phosphorus are displaced downstream as they complete a cycle (Webster and Patten 1979) through the generalized compartments of water, particulates, and consumers (Newbold et al. 1981). Spiraling length refers to the downstream distance required for one complete cycle or, for organic carbon, as the distance between its entry into the river corridor and its oxidation (Fig. 2.8) (Elwood et al. 1980). Spiraling length reflects the utilization of nutrients relative to the available supply, as well as physical characteristics of the river corridor, such as the ability to at least temporarily retain solutes and particulates in areas of reduced transport capacity (Fisher et al. 1998; Battin et al. 2008; Baker et al. 2012). Short spiraling lengths reflect high rates of material cycling, but disturbance can cause the spiraling length to increase (Fisher et al. 1998).

Lateral connectivity within river corridors is the focus of the flood-pulse model (Junk et al. 1989), which describes the ecological influence of the seasonal flood pulse on large floodplain rivers such as the Amazon (Fig. 2.9). Water, sediment, nutrients, and organisms move from the channel onto the floodplain during peak flood flow and then return to the main channel and secondary channels during the receding limb and base flow. This repeated movement enhances nutrient availability, as well as habitat and biodiversity within the channel and the floodplain. A subsequent iteration of this model emphasizes flow pulses, which are smaller-scale fluctuations in discharge that change the extent of flow and standing water within a braided or anabranching channel segment, as well as flow levels along the margins

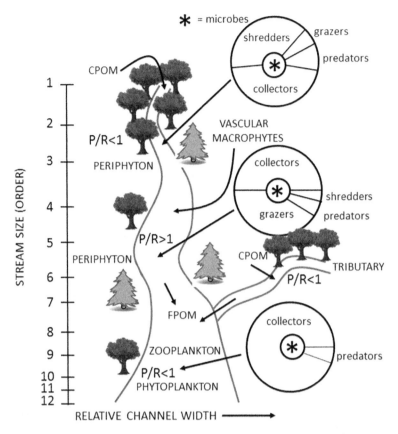

Fig. 2.7 Schematic illustration of the river continuum concept, which emphasizes progressive downstream trends in channel size, primary production (P), respiration (R), and functional groups of stream macroinvertebrates. Pie charts indicate relative proportions of functional groups. CPOM is coarse particulate organic matter. FPOM is fine particulate organic matter. P/R < 1 indicates that respiration exceeds primary production. Periphyton is biofilm. (After Vannote et al. 1980, Fig. 1)

of bars and islands within channels. Flow pulses enhance nutrient availability, habitat, and biodiversity (Tockner et al. 2000).

Changes through time have been conceptualized in terms of river resistance and resilience, as described previously, and in terms of alternate states. Alternate stable states describe ecological systems than can exist in multiple, distinct, and self-reinforcing states in equilibrium under equivalent environmental conditions (Holling 1973; May 1977). This scenario relies on a disturbance or a threshold in response to ongoing changing conditions that is capable of driving the ecosystem into an alternate state. The ecosystem then takes a different pathway of recovery in response to the disturbance, which ultimately leads to a self-sustaining reorganization of ecosystem structure. The existence of alternate stable states has typically been evaluated based on biotic community structure (e.g., Scheffer and Carpenter 2003), although the concept is controversial because of the difficulty in demonstrating

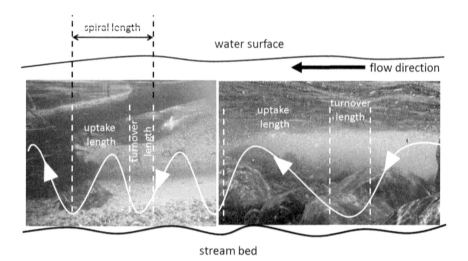

Fig. 2.8 Schematic illustration of nutrient spiraling length in river ecosystems. Nutrients spiral between particulate organic matter, including microbes, and the water column via uptake and turnover as the nutrients move downstream. Open spiraling, as in the right side of the illustration, means that nutrients remain in place for shorter periods of time. Tight spiraling, as in the left side of the illustration, means that nutrients remain in place longer. The background photo on the left is an underwater view of a logjam, which is one of the types of physical features that can help to retain nutrients in a stream. Uptake length describes the average distance traveled by dissolved nutrients in the water column before uptake. Turnover represents the average distance traveled by nutrients within biota, such as biofilms. The spiral length is the sum of the uptake and turnover lengths

long-term stability (Schröder et al. 2005). This has led to the related concept of alternate states, which can be stable or transient over varying time scales (Suding et al. 2004).

A well-known example of alternate states in river corridors is that of beaver meadows versus elk grasslands. Beaver meadows are the valley-bottom mosaic of beaver dams and ponds in varying stages of active maintenance or abandonment by beaver. Beaver meadows commonly include anabranching channels, ponds and marshes, and willow (*Salix* spp.) carrs (Polvi and Wohl 2012). An elk grassland has a drier valley bottom dominated by grasses rather than woody riparian vegetation, and typically has a single, incised channel with limited or no floodplain wetlands. A beaver meadow can become an elk grassland if the number of elk grazing in the riparian zone becomes large enough to suppress woody riparian species so that food supplies become insufficient to support beaver (Wolf et al. 2007). Reintroduction of elk predators such as wolves in river corridors where the predators were previously hunted to extinction can reduce elk herbivory on riparian vegetation. This allows beaver to recolonize the river corridor, which can return to the wetter, more spatially heterogeneous conditions of a beaver meadow (Ripple and Beschta 2004; Beschta and Ripple 2012).

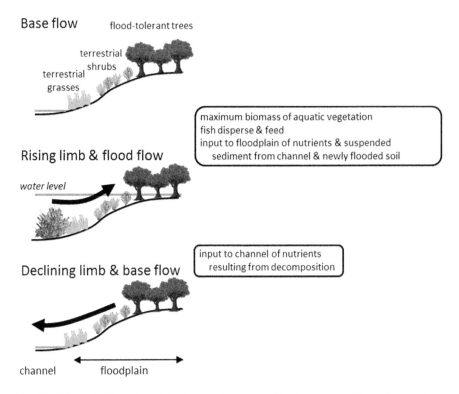

Fig. 2.9 Schematic illustration of the flood-pulse cycle and the importance of lateral connectivity within river corridors. (After Junk et al. 1989, Fig. 2)

As with conceptual models of physical process and form, conceptual models of ecological process and form emphasize particular aspects of a river ecosystem, such as longitudinal patterns, lateral connectivity, or alternations through time. These conceptual models can facilitate identification of particular processes and forms that are vital to river health, such as the flow pulses that temporarily enhance connectivity and nutrient processing in a braided river.

2.4.3   Commonalities of Physical and Ecological Conceptual Models

Areas of substantial overlap between physical and ecological conceptual models of river corridors include an emphasis on downstream trends and changes through time. Downstream trends can be conceptualized as primarily progressive downstream changes (downstream hydraulic geometry, river continuum concept) or abrupt downstream changes (process domains, river styles, serial discontinuity concept). Physical and ecological conceptualizations of change through time focus on the resistance of the river corridor to change and the resilience or characteristic form time of the river corridor following change. Geomorphology and ecology explicitly recognize the existence of thresholds and feedbacks, and the potential for a very different, persistent configuration of the river corridor following change (river metamorphosis, alternate state). The vocabulary and the illustrative examples differ between geomorphology and ecology, but the conceptual frameworks exhibit striking similarities that can be used to guide understanding and management of river health.

2.5 The 4Cs of River Health

What I refer to as the 4Cs of river health—connectivity, complexity, change, and context—is not an established phrase, but rather a useful way to remember some of the key factors that promote physical and ecological integrity. Physical integrity for rivers refers to a set of active river processes and landforms such that the river corridor adjusts to changes in water and sediment inputs within limits of change defined by societal values (Graf 2001). Ecological integrity describes the ability of the river corridor to support and maintain a community of organisms with species composition, diversity, and functional organization similar to those within natural habitats in the same region (Parrish et al. 2003). The definition of physical integrity focuses on a river corridor's ability to adjust to changing inputs. The definition of ecological integrity focuses on the river ecosystem's ability to sustain a relatively natural biotic community, which partly depends on the physical integrity of the river corridor. River integrity combines these definitions to describe the ability of the river ecosystem to adjust to changing water and sediment inputs (without constraints imposed by human manipulation such as dams or levees) and through these adjustments to maintain the habitat, disturbance regime, and connectivity necessary to sustain native biotic communities.

As noted briefly in the first chapter, determining natural conditions, whether physical or biotic, can be very difficult in a region with a long history of human occupation and resource use. Natural conditions are most commonly described in terms of reference conditions and natural range of variability. Reference conditions describe the characteristics of a natural system prior to intensive human alteration of that system. Because natural systems continually adjust through time, the range of adjustments is referred to as the natural (or sometimes the historical) range of variability (Keane et al. 2009). For a river corridor, the natural range of flow variability, for example, describes the upper and lower magnitude limits of discharge, as well as the range of duration, frequency, rates of change, and seasonal timing of various components of the hydrograph in the absence of human alteration of the watershed or manipulation of the flow (Poff et al. 1997) (Fig. 2.10).

Two particularly common forms of information on reference conditions and natural range of variability are historical data from the site of interest and contemporary data from reference sites that are chosen as appropriate analogs for the site of

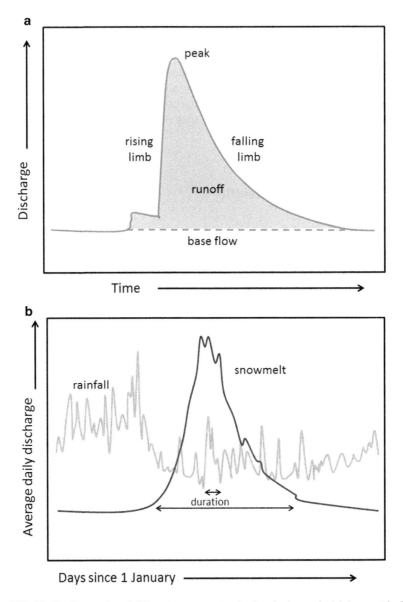

Fig. 2.10 Idealized examples of different components of a river hydrograph. (**a**) An event hydrograph for a single flood, showing the rising limb, flood peak, falling limb, and volume of storm runoff that exceeds base flow. (**b**) Average annual hydrograph for a river dominated by snowmelt runoff, which has a single annual peak of longer duration and slower rates of rise and fall than the more frequent but shorter duration peaks of a river dominated by rainfall runoff

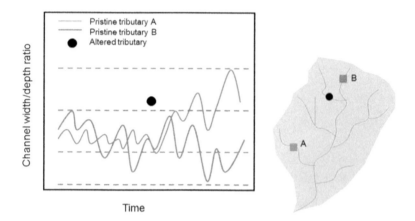

Fig. 2.11 Schematic illustration of natural range of variability. The *black circle* indicates the width/depth ratio of a river segment on an altered tributary. The *blue line* represents variations through time in the width/depth ratio of pristine tributary A: the natural range of variability is indicated by the *dashed blue lines* defining the upper and lower limits of the width/depth ratio for this site. The *brown lines* represent similar information for pristine tributary B. If the altered tributary is compared only to site A, it lies within the natural range of variability. Comparison to site B suggests that the altered tributary is outside of the natural range of variability

interest (White and Walker 1997). Historical data can be problematic because of unmeasured factors that confound the interpretation of observed historical changes. An analysis designed to relate changes in channel geometry to precipitation and stream flow in a forested catchment, for example, might misinterpret a change in channel geometry associated with a widespread blowdown that recruited substantial large wood to the channel network. Inferences based on reference sites can be problematic because of the difficulty in finding a close match in all relevant parameters (White and Walker 1997). Use of either historical data or reference sites can be constrained by limited information (Fig. 2.11), but comparison to relatively natural river corridors remains the most widely used method of assessing the effects of human alterations on river ecosystems.

Despite the difficulties of characterizing natural range of variability and reference conditions, attention to four primary aspects of river corridors as part of river management can enhance and sustain river health. The following sections explore each aspect in more detail.

2.5.1   Connectivity

Connectivity describes movements of energy, material, and organisms between components of a system, which are typically spatially defined areas. Structural connectivity (Wainwright et al. 2011) results from physical contiguity (e.g., water flows from a hillslope directly onto a floodplain). Functional connectivity results from

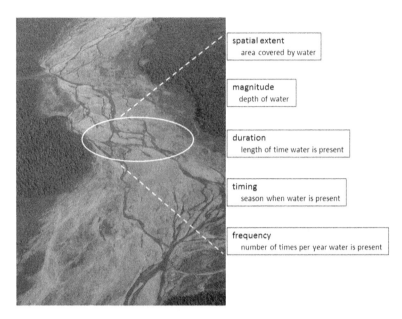

Fig. 2.12   An illustration of different characteristics of connectivity for a section of river corridor in northern Alaska. In this braided river segment, the presence and amount of surface water at any point within the river corridor vary through time. If connectivity is characterized as the presence of surface water, the details of this connectivity can be described in terms of spatial extent, magnitude, duration, timing, and frequency

transport between components of a system that may not be physically contiguous (e.g., wind carries silt and clay eroded from the Gobi Desert across the Pacific Ocean and deposits the material on a channel in western North America). Connectivity can be characterized in terms of spatial extent, magnitude, duration, timing, and frequency (Fig. 2.12). Connectivity also implies the presence of disconnectivity, as when an erosionally resistant bedrock unit, for example, limits upstream transmission of base level fall (Fig. 2.13) or when insufficient surface flow causes segments of a river to become dry (Fig. 2.14).

A river corridor has three primary dimensions of connectivity both within the corridor and between the corridor and adjacent environments (Ward 1989; Pringle 2001) (Fig. 2.15). Longitudinal connectivity is perhaps the most obvious dimension. Water, solutes, sediment, organic matter, and organisms move downstream, and some organisms move upstream within the river network (Vannote et al. 1980; Cote et al. 2009). Lateral connectivity can be bi-directional within the river corridor as high flows spread from the main channel into secondary channels and across the floodplain, and then recede back to the main channel (Junk et al. 1989). Lateral connectivity can also be unidirectional as materials and organisms move from uplands into the river corridor. Vertical connectivity is also bidirectional above and below the river and floodplain. Water, volatile chemicals, and organisms move upward into the atmosphere (Ward 1989). Wet and dry atmospheric deposition, as well as ter-

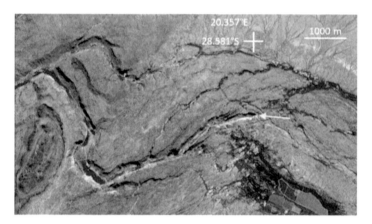

Fig. 2.13 Aerial view of the Orange River at Augrabies Falls in South Africa. The 60-m-tall waterfall limits upstream transmission of base level fall along the river. *Yellow arrow* at head of falls indicates flow direction. (Photograph courtesy of Google Earth)

restrial insects, move downward into the river corridor. Water, solutes, microbes, and macroinvertebrates move vertically between the channel and the hyporheic zone, and water and associated solutes move vertically between the river corridor and the ground water (Brunke and Gonser 1997).

Thinking about a river corridor or river network in the context of connectivity is particularly useful because this viewpoint emphasizes fluxes and the processes that maintain these fluxes. This conceptualization also promotes attention to aspects of rivers that are more likely to be neglected, including the hyporheic zone, the floodplain, and headwater channels. The hyporheic zone may be neglected because it is largely invisible to anyone viewing the river corridor from the ground surface or above. The floodplain may be neglected because it is commonly viewed as somehow detached from or irrelevant to the active channel in regulatory contexts. Headwater channels can be neglected because they are too small to support fisheries or navigation and because they are likely to be viewed as unimportant to the functioning and integrity of river networks (e.g., in regulatory contexts, which seldom protect very small channels; Meyer and Wallace 2001; Meyer et al. 2007; Nadeau and Rains 2007). Each of these sometimes-neglected components of hyporheic zone, floodplain, and headwater channels is in reality vital to the functioning of a river ecosystem.

Connectivity is not necessarily good or bad in the context of river integrity: the details matter. A beaver dam that reduces longitudinal connectivity also increases lateral and vertical connectivity, creating enhanced storage of water, sediment, and organic matter, as well as increased habitat and biodiversity. A waterfall that reduces longitudinal connectivity for upstream fish migration also reduces longitudinal connectivity for exotic fish species that might otherwise colonize portions of the river network above the waterfall. The key point is that river integrity depends on a natu-

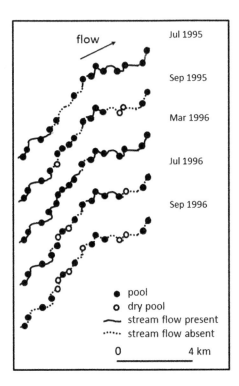

Fig. 2.14 Illustration of changing longitudinal connectivity over time spans of a few months along an intermittent channel in eastern Colorado, USA. Pools that provide habitat for endangered native fish species can become dry. A reduction in the number of pools limits habitat for the fish. Absence of surface flow between the pools limits dispersal by fish, which limits genetic exchange and long-term viability of the fish population, as well as resulting in death of all the fish within a pool when it goes dry. Consequently, the temporal sequence and spatial patterns of drying in pools and intervening channel segments are very important to the survival of native fish species. (Modified from Fausch et al. 2002, Fig. 8)

ral range of variability for connectivity, which in some river networks may mean that sources of disconnectivity are vital to river health.

Natural forms of disconnectivity are rarely permanent or complete, however, and are better conceptualized as creating temporary limits to connectivity. A tributary alluvial fan that stores sediment transported by a tributary channel to the main river corridor, for example, eventually releases that sediment for transport down the main stem river, although a period of centuries to millennia may pass before the main stem remobilizes the tributary sediment (Meyer et al. 1992). The tributary fan may also slow the passage of water that infiltrates from the tributary channel and moves into the main stem river corridor over a period of weeks to a few years, helping to sustain base flows and riparian water tables in the main river corridor (Woods et al. 2006). A waterfall that limits upstream migration of fish or amphibians may persist for thousands of years, allowing geographic speciation in isolated populations of river organisms (May et al. 2017), but erosion will remove the waterfall over geo-

Fig. 2.15 Schematic illustration of the three primary dimensions of connectivity: longitudinal, lateral, and vertical, here superimposed on an upstream view of East Inlet Creek in Rocky Mountain National Park, Colorado, USA

logical timescales of tens of thousands to millions of years. Natural forms of disconnectivity can thus be persistent, impassable barriers over timespans relevant to human and biotic communities, but can also effectively slow rather than completely stop connectivity.

In summary, understanding and maintaining river health relies on understanding the processes that create and maintain natural levels of connectivity and being able to characterize natural levels of connectivity.

2.5.2 Complexity

Complexity has at least two meanings within river science. The first refers to complex behavior and the presence of nonlinear dynamics, self-organization, and emergent properties (Werner and McNamara 2007; Phillips 2014). The second uses complexity to describe spatial heterogeneity, or the degree to which a river corridor deviates from a straight, uniform canal. Complex behavior and spatial heterogeneity are commonly interrelated (Phillips 2007; Peipoch et al. 2015), but this discussion focuses on spatial heterogeneity.

Spatial heterogeneity in a river corridor occurs at diverse spatial scales from a channel unit or patch of sediment up to the entire longitudinal profile or river network (Fig. 2.16). At any of these scales, heterogeneity results from discrete patches with differing characteristics. These patches can be described in terms of patch richness (the number of patches present), patch frequency (the number of each patch

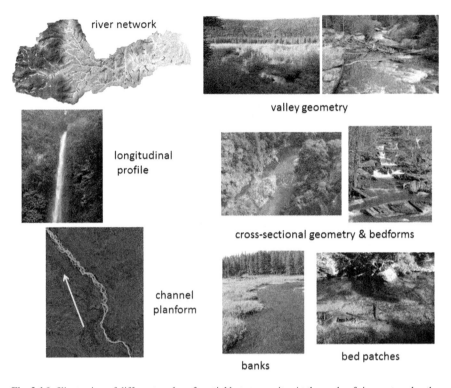

Fig. 2.16 Illustration of different scales of spatial heterogeneity. At the scale of river networks, the inset map of the North St. Vrain Creek drainage in Colorado, USA (*top left*) has differing channel gradients throughout the network, as indicated by different line colors in the map. At the scale of the longitudinal profile, the photo at middle left shows a waterfall that separates longitudinal segments of differing gradient along a river in Costa Rica. At the scale of the channel planform, the photo at bottom left shows how the Yukon River in central Alaska varies from straight to anabranching proceeding downstream (*yellow arrow* indicates flow direction). At the scale of valley geometry, the two photos at upper right illustrate how some reaches of North St. Vrain Creek have wide, low gradient valleys with substantial channel-floodplain connectivity (*left photo*), whereas others have steep, narrow valleys with limited lateral connectivity (*right photo*). At the scale of cross-sectional geometry and bedforms, the photos in the middle right show a pool (*top*) and riffle (*center*) along the Rio Chagres in Panama in the left photo and a sequence of steps and pools along a small channel in Tennessee, USA in the right photo. At the scale of channel banks and bed, the left photo in the pair of photos at lower right illustrates sub-meter-scale undulations along the bank of Hog Park Creek in Wyoming, USA. The right photo in the pair shows grain-size patches formed by boulders (*top*) and sand and organic matter (*bottom*)

type), patch configuration (patch arrangement in space relative to each other), patch change (which patches change and how they change), and a shifting mosaic (patch changes are spatially explicit and quantified) (Cadenasso et al. 2006). River scientists have devoted substantial effort to quantifying these aspects of complexity using spatial statistics and other metrics and there are now dozens of numerical descriptors that can be applied to particular aspects of complexity in river corridors (Wohl 2016).

Complexity is important in the context of river integrity or river health for at least six reasons.

1. Complexity in the form of habitat diversity can correlate with biodiversity. Although other factors such as introduced species or limited connectivity can constrain biodiversity, several studies find that greater habitat diversity in river corridors correlates with greater biodiversity and biological productivity (Scott et al. 2003; Luck et al. 2010; Bellmore and Baxter 2014; Greene and Knox 2014).

2. Complexity influences attenuation of downstream fluxes. Diverse forms of complexity—from bedforms and instream wood to extensive floodplains with varying surface topography and vegetation communities—increase hydraulic resistance, reduce average downstream velocity, enhance hyporheic exchange, and facilitate at least transient deposition of organic matter and sediment (Jeffries et al. 2003; Brummer et al. 2006; Ensign and Doyle 2006; Westbrook et al. 2006; Gooseff et al. 2007; Small et al. 2008; Wohl and Scott 2017). If organic matter is stored even for a short period of a few minutes to hours, it becomes more available to aquatic and riparian organisms (Battin et al. 2008) and thus helps to support greater biological productivity within the river corridor.

3. Complexity influences the resistance and resilience of river corridors to disturbance. A more complex river corridor can be more resistant and resilient to disturbance, as illustrated by a river corridor in which numerous channel-spanning logjams promote formation of an anabranching channel (Wohl 2011; Collins et al. 2012). During peak flows, flood waters spread among the multiple channels and across the floodplain. Velocity and flow energy are reduced because of the relatively shallow flow depths and large hydraulic resistance, so the river corridor is more resistant to the flood disturbance than it would be in the absence of the complexity associated with the logjams. The spatial heterogeneity of the channel boundaries creates multiple low-velocity refuges for fish and other organisms during the flood, allowing rapid recolonization of habitats throughout the river corridor following the flood and thus increasing the resilience of the river corridor to disturbance.

4. Complexity both reflects and influences processes in rivers. The spatial heterogeneity present in the channel, floodplain, and hyporheic zone each reflect the combined history of hydraulic force, sediment movement, and biotic colonization (vegetation growth, beaver dams) through time. Habitat diversity, for example, strongly correlates with channel mobility (Choné and Biron 2016). Complexity influences river processes when bedforms or instream wood increase hydraulic resistance, which in turn influences hydraulic force and sediment movement (e.g., Yochum et al. 2012), or when logjams abandoned during channel migration and then buried in the floodplain form erosionally resistant points that limit subsequent channel migration (Collins et al. 2012).

5. Enhancing complexity is increasingly an explicit goal of river management. Most of the history of river management has involved making river corridors simpler and more homogeneous, as discussed in the next chapter. This simplifi-

cation and homogenization has resulted in widespread negative consequences, however, from loss of flood attenuation and nutrient uptake to sharp declines in biodiversity. Consequently, river managers and society as a whole are emphasizing the need to restore physical complexity. Restoration can ta ke the form of engineering channel and floodplain environments to be more spatially heterogeneous, such as adding sinuosity to channels or introducing large boulders and instream wood to channels (Nilsson et al. 2005; Helfield et al. 2007; Polvi et al. 2014). Restoration of physical complexity can also involve restoring processes that create and maintain spatial heterogeneity, such as reducing flow regulation in order to create more natural water and sediment regimes (Poff et al. 1997; Wohl et al. 2015).

6. Complexity reflects the inherent unevenness through time and space of processes within a drainage basin and river corridor. This unevenness is reflected in biogeochemical reactions by the existence of hot spots and hot moments. Hot spots are patches with disproportionately high reaction rates relative to the surrounding matrix. Hot moments refer to short periods with disproportionately high reaction rates relative to longer intervening time periods (McClain et al. 2003). Unevenness is reflected in sediment dynamics by numerous examples of drainages in which a small portion of the catchment contributes most of the sediment (e.g., Meade et al. 1985; Meade 2007) and the great majority of sediment flux occurs during a very short period of time (e.g., Farnsworth and Milliman 2003; Kao and Milliman 2008). Unevenness is represented by organic carbon storage in river networks in which limited portions of the network store the majority of carbon (e.g., Wohl et al. 2012).

In summary, diverse forms of complexity drive physical processes, biogeochemical processes, habitat diversity, and biodiversity in river ecosystems. Consequently, understanding the types of complexity present in a river ecosystem, as well as the factors that create and maintain the complexity, are critical to understanding and maintaining river health.

2.5.3   *Change*

Changes through space and time are inherent in natural river corridors. Tributaries differ from the main channel. Valley geometry and disturbance regime vary longitudinally. Disturbances such as seasonal floods occur at predictable intervals, but other disturbances such as drought or forest blowdown occur at longer, more irregular intervals. Heredotus' famous phrase that you can never step in the same river twice incorporates many layers of meaning.

Both spatial and temporal changes are critical to river integrity because they create and maintain habitat diversity. Human societies tend to regard large floods as disasters, for example, but these high-energy flows are critical sources of complexity in river corridors. The big floods promote channel avulsion and the formation of secondary or cutoff channels, as well as creating new germination sites for riparian vegetation and redistributing nutrients across the river corridor. Big floods recruit large wood to the channel and floodplain and redistribute that wood into jams that provide critical habitat for aquatic and terrestrial species. Big floods mobilize the bed in coarse-grained channels, flushing interstitial fine sediment and maintaining hyporheic exchange and fish spawning habitat. From a river ecosystem perspective, big floods are vitally important to river health.

A critical corollary of change is the capacity of the river to adjust to disturbances. Returning to the definition of physical integrity, the location, dimensions, and configuration of a river channel must be able to adjust to changing water and sediment inputs. The organisms within the river corridor must be able to seek refuge during disturbance and to migrate and recolonize following disturbance. Consequently, the continuing occurrence of changes and the capacity to adjust to these changes characterize river integrity. Understanding the processes that drive change, as well as the magnitude and frequency of these processes, is thus critical to understanding and maintaining river health.

2.5.4 *Context*

The final C of river health is context, which here refers to the environmental setting of the river corridor: the climate and precipitation regime, geology, topography, biome, position in the river network, land use, and land use history. The physics and chemistry underlying river process and form are consistent around the world, but the specific details of local history as expressed through climate, geology, evolution, and human history can differ enormously and these details matter to river integrity and river management. Numerous classification systems have been proposed for rivers in recognition of these differences. This section focuses on three that are applied at relatively broad scales within or between river networks: ecoregions, process domains, and river styles.

Ecologists designate terrestrial and freshwater ecoregions based on characteristic combinations of physical factors (topography, soils) and biotic communities (Omernik 2004). An ecoregion is a relatively large area of land or water that contains a geographically distinct assemblage of natural communities (WWF 2013). Bailey (1995), for example, describes 25 divisions of terrestrial ecoregions that encompass 54 provinces within the United States. The World Wildlife Fund designates 76 freshwater ecoregions within North America (Fig. 2.17), about 45 of which are within the United States (WWF 2013). Either terrestrial or freshwater ecoregions provide a broad-scale differentiation of common characteristics and a starting point when seeking reference sites for river ecosystems. Unfortunately, only 17 of the freshwater ecoregions in North America were rated as being relatively stable or relatively intact, in terms of overall conservation status, in 2013 (WWF 2013). These ecoregions in good condition are concentrated in Alaska and northern Canada. The remainder of the ecoregions were rated as being critical, endangered, or vulner-

Fig. 2.17 Map of freshwater ecoregions in North America. (After WWF 2013, Fig. 2.1, p. 10)

able based on characteristics such as alteration of land cover, surface water quality, alteration of flow regime, degree of habitat fragmentation, introduced species, and imperiled native species (WWF 2013).

Process domains are spatially discrete portions of a landscape or river network characterized by distinct suites of geomorphic processes (Montgomery 1999). Within a drainage network, individual process domains might be differentiated based on disturbance regime. Within a mountainous river network, for example, one process domain at higher elevations might be defined based on the occurrence of snowmelt-runoff flow peaks, whereas a second process domain might be defined at lower elevations based on rainfall-runoff flow peaks. Process domains can also be distinguished based on how river segments resist or recover following disturbance. Considering a mountainous river network again, one process domain could consist of steep, narrow valley segments with erosionally resistant substrate such as bedrock and large boulders and channel geometry predominantly formed during large flows. Another process domain could consist of lower gradient, wider valley segments with mobile alluvium in which channel geometry reflects predominantly more frequent flows of lower magnitude (Fig. 2.18). Regardless of the specific criteria used to delineate process domains, they are likely to reflect finer spatial scales than ecoregions.

River style (Brierley and Fryirs 2005) is a geomorphic classification applied at the reach scale that is similar to process domains in recognizing spatially discrete reaches that are internally homogeneous with respect to topography, valley geometry, channel planform, and substrate, but differ from other reaches (Fig. 2.19). There is substantial overlap between the conceptualization of process domains and river

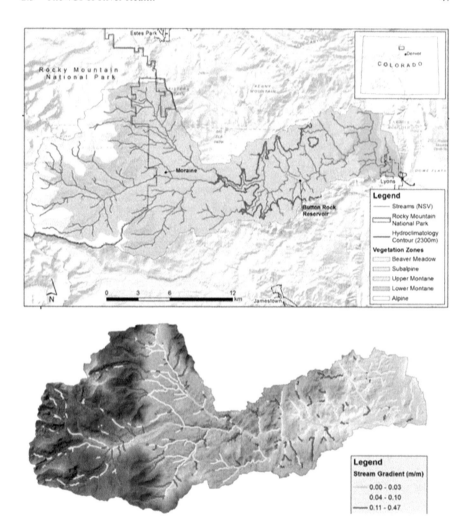

Fig. 2.18 Schematic illustration of two means of distinguishing process domains, using the North St. Vrain Creek watershed (drainage area 340 km²) in Colorado, USA as an example. In the upper map, process domains are differentiated based on disturbance regime. The alpine and subalpine portions of the drainage experience only snowmelt floods, whereas the montane portions also experience floods from convective rainfall. Stand-killing fires and associated debris flows are much more common within the montane zone than in the alpine and subalpine zones. In the lower map, segments of differing channel gradient represent process domains. High-gradient segments have narrow valley bottoms and lower gradient segments have wider valley bottoms

styles, although the river styles framework provides more explicit guidelines on the criteria used to delineate river styles.

Use of ecoregions, process domains, or river styles reflects recognition that context matters. Distinct portions of an individual river network, as well as river networks in different regions of the world, will differ in terms of process, form,

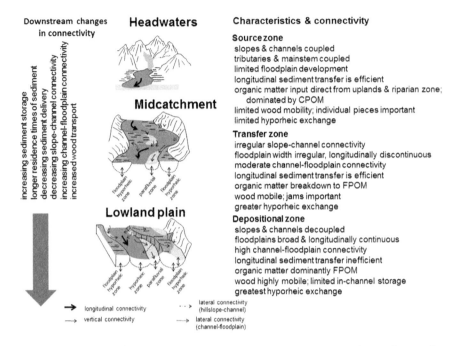

Fig. 2.19 An illustration of three river styles, including differences in diverse forms of connectivity. (After Brierley and Fryirs 2005, Fig. 2.10)

connectivity, complexity, disturbance regime, biota, and many other factors. Understanding and maintaining river health depends on being able to recognize these natural differences, rather than assuming that a management template good for one river segment or network is equally good for all river segments or networks (Brierley and Fryirs 2009).

2.6 River Ecosystem Services and Water Resources

What are the implications of viewing river corridors as ecosystems rather than simple conveyance or water supply channels? The most important implication is likely to be that rivers conceptualized as ecosystems will be managed differently than rivers have been under traditional practices. Traditional attitudes toward water resources emphasize human consumptive uses and river engineering designed to facilitate those uses. Premier among consumptive uses, in terms of long history and the single greatest consumption of water at regional to global scales, is irrigated agriculture. Within watersheds that include irrigated agriculture, and globally, irrigated agriculture accounts for 80–90% of consumptive water use (Scanlon et al. 2007). Some of this water comes from ground water pumping, but much of it comes from surface water in rivers and lakes. River corridors have been extensively modified to ensure that crop lands receive irrigation water when and where it is needed and to ensure that water tables remain at levels necessary to sustain crops. Other critically important components of water resources are household, municipal, and industrial water supplies; wastewater treatment; river navigation; and flood control. Recreational uses and subsistence fisheries, although economically and societally important in some regions, generally receive less emphasis in water resources management.

Simply stated, traditional water resources management has striven to (1) make river corridors as physically simple and uniform as possible in order to ensure downstream conveyance of flood waters and wastewater and to maintain navigation and (2) control the downstream movement of water and sediment to ensure a steady supply of water when needed and to stabilize channels and floodplains in a static configuration. To these ends, the great majority of river corridors in the temperate latitudes have been highly engineered via dredging, channelization, levees, flow regulation, and land drainage. Consequently, there now remain almost no natural river corridors around the world except at higher or lower latitudes. Despite the enormous and ever-increasing cost of river and water resources engineering, one can argue that this engineering has been successful in supporting the 174% expansion of irrigated agriculture that has occurred globally since the 1950s (Scanlon et al. 2007) and the associated increase in global population from 2.5 billion in 1950 to 6.5 billion in 2005 (Bongaarts 2009). One can also argue that this success has come at unsustainable cost to freshwater and nearshore ecosystems and that rates of increase in global standard of living cannot be maintained unless much greater emphasis is given to protecting and restoring river ecosystem services.

So how exactly can river engineering and management be altered to better sustain rivers as ecosystems that provide vital ecosystem services? This topic is the focus of Chap. 4, but the short answer is by restoring more natural processes—spatial and temporal variability in water and sediment inputs—that in turn support more spatially heterogeneous and changeable river corridors, with sufficient habitat diversity and connectivity to support diverse biotic communities. Before exploring this new vision of management in more detail, however, Chap. 3 provides a more in-depth review of the history and types of human activities that have altered river ecosystems.

References

Abbe TB, Montgomery DR (2003) Patterns and processes of wood debris accumulation in the Queets River basin, Washington. Geomorphology 51:81–107

Arnaud F, Piegay H, Schmitt L, Rollet AJ, Ferrier V, Beal D (2015) Historical geomorphic analysis (1932–2011) of a by-passed river reach in process-based restoration perspectives: the Old Rhine downstream of the Kembs diversion dam (France, Germany). Geomorphology 236:163–177

Baartman JEM, Masselink R, Keesstra SD, Temme AJAM (2013) Linking landscape morphological complexity and sediment connectivity. Earth Surf Process Landf 38:1457–1471

Bailey RG (1995) Description of the ecoregions of the United States, 2nd edn. USDA Forest Service Miscellaneous Publication 1391, Washington, DC

Baker DW, Bledsoe BP, Price JM (2012) Stream nitrate uptake and transient storage over a gradient of geomorphic complexity, north-central Colorado, USA. Hydrol Process 26:3241–3252

Baker VR (1988) Flood erosion. In: Baker VR, Kochel RC, Patton PC (eds) Flood geomorphology. Wiley, New York, pp 81–95

Battin TJ, Kaplan LA, Findlay S, Hopkinson CS, Marti E, Packman AI, Newbold JD, Sabater F (2008) Biophysical controls on organic carbon fluxes in fluvial networks. Nat Geosci 1:95–100

Battin TJ, Besemer K, Bengtsson MM, Romani AM, Packman AI (2016) The ecology and biogeochemistry of stream biofilms. Nat Rev Microbiol 14:251–263

Baxter CV, Fausch KD, Saunders WC (2005) Tangled webs: reciprocal flows of invertebrate prey link streams and riparian zones. Freshw Biol 50:201–220

Bayley PB (1991) The flood pulse advantage and the restoration of river-floodplain systems. River Res Appl 6:75–86

Beckman ND, Wohl E (2014) Carbon storage in mountainous headwater streams: the role of old-growth forest and logjams. Water Resour Res 50:2376–2393

Bellmore JR, Baxter CV (2014) Effects of geomorphic process domains on river ecosystems: a comparison of floodplain and confined valley segments. River Res Appl 30:617–630

Benda L (1990) The influence of debris flows on channels and valley floors in the Oregon Coast Range, USA. Earth Surf Process Landf 15:457–466

Benda LE, Sias JC (2003) A quantitative framework for evaluating the mass balance of in-stream organic debris. For Ecol Manag 172:1–16

Beschta RL, Ripple WJ (2012) The role of large predators in maintaining riparian plant communities and river morphology. Geomorphology 157–158:88–98

Boivin M, Buffin-Bélanger T, Piégay H (2015) The raft of the Saint-Jean River, Gaspé (Québec, Canada): a dynamic feature trapping most of the wood transported from the catchment. Geomorphology 231:270–280

Bongaarts J (2009) Human population growth and the demographic transition. Philos Trans R Soc B 364:2985–2990

Braccia A, Batzer DP (2001) Invertebrates associated with woody debris in a southeastern US forested floodplain wetland. Wetlands 21:18–31

Brierley GJ, Brooks AP, Fryirs K, Taylor MP (2005) Did humid-temperate rivers in the Old and New Worlds respond differently to clearance of riparian vegetation and removal of woody debris? Prog Phys Geogr 29:27–49

Brierley GJ, Fryirs KA (2005) Geomorphology and river management: applications of the river styles framework. Blackwell, Oxford, p 398

Brierley G, Fryirs K (2009) Don't fight the site: three geomorphic considerations in catchment-scale river rehabilitation planning. Environ Manag 43:1201–1218

Brummer CJ, Abbe TB, Sampson JR et al (2006) Influence of vertical channel change associated with wood accumulations on delineating channel migration zones, Washington, USA. Geomorphology 80:295–309

Brunke M, Gonser T (1997) The ecological significance of exchange processes between rivers and groundwater. Freshw Biol 37:1–33

Brunsden D, Thornes JB (1979) Landscape sensitivity and change. Trans Inst Br Geogr 4:463–484

Buffington JM, Montgomery DR (1999) Effects of hydraulic roughness on surface textures of gravel-bed rivers. Water Resour Res 35:3507–3521

Buffington JM, Tonina D (2009) Hyporheic exchange in mountain rivers II: effects of channel morphology on mechanics, scales, and rates of exchange. Geogr Compass 3:1038–1062

Bull WB (1991) Geomorphic responses to climatic change. Oxford University Press, New York

Burchsted D, Daniels M, Thorson R, Vokoun J (2010) The river discontinuum: applying beaver modifications to baseline conditions for restoration of forested headwaters. Bioscience 60:908–922

Cadenasso ML, Pickett STA, Grove JM (2006) Dimensions of ecosystem complexity: heterogeneity, connectivity, and history. Ecol Complex 3:1–12

Cenderelli DA, Wohl EE (2003) Flow hydraulics and geomorphic effects of glacial-lake outburst floods in the Mount Everest region, Nepal. Earth Surf Process Landf 28:385–407

Choné G, Biron PM (2016) Assessing the relationship between river mobility and habitat. River Res Appl 32:528–539

Collins BD, Montgomery DR, Fetherston KL, Abbe TB (2012) The floodplain large-wood cycle hypothesis: a mechanism for the physical and biotic structuring of temperate forested alluvial valleys in the North Pacific coastal ecoregion. Geomorphology 139–140:460–470

Constantine JA, Dunne T, Ahmed J, Legleiter C, Lazarus ED (2014) Sediment supply as a driver of river meandering and floodplain evolution in the Amazon Basin. Nat Geosci 7:899–903

Cote D, Kehler DG, Bourne C, Wiersma YF (2009) A new measure of longitudinal connectivity for stream networks. Landsc Ecol 24:101–113

Cuffney TF (1988) Input, movement, and exchange of organic matter within a subtropical coastal black water river-floodplain system. Freshw Biol 19:305–320

Curran JH, Wohl EE (2003) Large woody debris and flow resistance in step-pool channels, Cascade Range, Washington. Geomorphology 51:141–157

Dunkerley D (2014) Nature and hydro-geomorphic roles of trees and woody debris in a dryland ephemeral stream: Fowlers Creek, arid western New South Wales, Australia. J Arid Environ 102:40–49

Dunne T, Aalto RE (2013) Large river floodplains. In: Wohl E, Shroder JF (eds) Fluvial geomorphology. Treatise on geomorphology, vol 9. Elsevier, pp 645–677

Dunne T, Mertes LAK, Meade RH, Richey JE, Forsberg BR (1998) Exchanges of sediment between the flood plain and channel of the Amazon River in Brazil. Geol Soc Am Bull 110:450–467

Elwood JW, Newbold JD, O'Neill RV, Van Winkle W (1980) Resource spiralling: an operational paradigm for analyzing lotic ecosystems. Stream Ecology Symposium, Augusta, GA

Ensign SH, Doyle MW (2006) Nutrient spiraling in streams and river networks. J Geophys Res 111:G04009

Erskine W, Keene A, Bush R, Cheetham M, Chalmers A (2012) Influence of riparian vegetation on channel widening and subsequent contraction on a sand-bed stream since European settlement: Widden Brook, Australia. Geomorphology 147–148:102–114

Farnsworth KL, Milliman JD (2003) Effects of climatic and anthropogenic change on small mountainous rivers: the Salinas River example. Glob Planet Chang 39:53–64

Fausch KD, Torgersen CE, Baxter CV, Li HW (2002) Landscapes to riverscapes: bridging the gap between research and conservation of stream fishes. Bioscience 52:483–498

Fisher SG, Grimm NB, Marti E, Holmes RM, Jones JB (1998) Material spiraling in stream corridors: a telescoping ecosystem model. Ecosystems 1:19–34

Friedman JM, Lee VJ (2002) Extreme floods, channel change, and riparian forests along ephemeral streams. Ecol Monogr 72:409–425

Fryirs KA (2016) River sensitivity: a lost foundation concept in fluvial geomorphology. Earth Surf Process Landf 42:55–70

Fryirs KA, Brierley GJ, Preston NJ, Kasai M (2007) Buffers, barriers and blankets: the (dis)connectivity of catchment-scale sediment cascades. Catena 70:49–67

Gende SM, Edwards RT, Willson MF, Wipfli MS (2002) Pacific salmon in aquatic and terrestrial ecosystems. Bioscience 52:917–928

Gomez B, Eden DN, Peacock DH, Pinkney EJ (1998) Floodplain construction by recent, rapid vertical accretion: Waipaoa River, New Zealand. Earth Surf Process Landf 23:405–413

Gooseff MN (2010) Defining hyporheic zones—advancing our conceptual and operational definitions of where stream water and groundwater meet. Geogr Compass 4:945–955

Gooseff MN, Hall RO, Tank JL (2007) Relating transient storage to channel complexity in streams of varying land use in Jackson Hole, Wyoming. Water Resour Res 43:W01417

Graf WL (2001) Damage control: restoring the physical integrity of America's rivers. Ann Assoc Am Geogr 91:1–27

Greene SL, Knox JC (2014) Coupling legacy geomorphic surface facies to riparian vegetation: assessing red cedar invasion along the Missouri River downstream of Gavins Point Dam, South Dakota. Geomorphology 204:277–286

Griffin ER, Kean JW, Vincent KR, Smith JD, Friedman JM (2005) Modeling effects of bank friction and woody bank vegetation on channel flow and boundary shear stress in the Rio Puerco, New Mexico. J Geophys Res 110:F04023. doi:10.1029/2005JF000322

Gurnell AM (1998) The hydrogeomorphological effects of beaver dam-building activity. Prog Phys Geogr 22:167–189

Harvey AM (2012) The coupling status of alluvial fans and debris cones: a review and synthesis. Earth Surf Process Landf 37:64–76

Harvey JW, Fuller CC (1998) Effect of enhanced manganese oxidation in the hyporheic zone on basin-scale geochemical mass balance. Water Resour Res 34:623–636

Helfield JM, Capon SJ, Nilsson C, Jansson R, Palm D (2007) Restoration of rivers used for timber floating: effects on riparian plant diversity. Ecol Appl 17:840–851

Hester ET, Doyle MW (2008) In-stream geomorphic structures as drivers of hyporheic exchange. Water Resour Res 44:W03417. doi:10.1029/2006WR005810

Holling CS (1973) Resilience and stability of ecological systems. Ann Rev Ecol Syst 4:1–23

Hygelund B, Manga M (2003) Field measurements of drag coefficients for model large woody debris. Geomorphology 51:175–185

Jeffries R, Darby SE, Sear DA (2003) The influence of vegetation and organic debris on floodplain sediment dynamics: case study of a low-order stream in the New Forest, England. Geomorphology 51:61–80

John S, Klein A (2004) Hydrogeomorphic effects of beaver dams on floodplain morphology: avulsion processes and sediment fluxes in upland valley floors (Spessart, Germany). Quaternaire 15:219–231

Johnson LB, Breneman DH, Richards C (2003) Macroinvertebrate community structure and function associated with large wood in low gradient streams. River Res Appl 19:199–218

Johnston CA (2014) Beaver pond effects on carbon storage in soils. Geoderma 213:371–378

Junk WJ, Bayley PB, Sparks RE (1989) The flood-pulse concept in river-floodplain systems. Can Spec Publ Fish Aquat Sci 106:110–127

Kao SJ, Milliman JD (2008) Water and sediment discharge from small mountainous rivers, Taiwan: the roles of lithology, episodic events, and human activities. J Geol 116:431–448

Keane RE, Hessburg PF, Landres PB, Swanson FJ (2009) The use of historical range and variability (HRV) in landscape management. For Ecol Manag 258:1025–1037

Kieffer SW (1989) Geologic nozzles. Rev Geophys 27:3–38

Korup O (2013) Landslides in the fluvial system. In: Wohl E (ed) Treatise on fluvial geomorphology. Treatise on geomorphology, vol 9. Academic, San Diego, pp 244–259

Kramer N, Wohl E, Hess-Homeier B, Leisz S (2017) The pulse of driftwood export from a very large forested river basin over multiple time scales, Slave River, Canada. Water Resour Res 53:1928–1947

Kuhnle RA (2013) Suspended load. In: Wohl E (ed) Treatise on fluvial geomorphology. Treatise on geomorphology, vol 9. Academic, San Diego, pp 124–136

Laenen A, Risley JC (1997) Precipitation-runoff and streamflow-routing models for the Willamette River basin, Oregon. U.S. Geological Survey Water Resources Investigations Report 95-4284. 197 pp

Larsen A, May JH, Moss P, Hacker J (2016) Could alluvial knickpoint retreat rather than fire drive the loss of alluvial wet monsoon forest, tropical northern Australia? Earth Surf Process Landf 41:1583–1594

Lautz LK, Siegel DI (2007) The effect of transient storage on nitrate uptake lengths in streams: an inter-site comparison. Hydrol Process 21:3533–3548

Leopold LB, Maddock T (1953) The hydraulic geometry of stream channels and some physiographic implications. U.S. Geological Survey Professional Paper 252. Washington, DC, 56 pp

Lichvar RW, Wakeley JS (2004) Review of ordinary high water mark indicators for delineating arid streams in the southwestern United States. Technical Reports, U.S. Army Engineer Waterways Experiment Station, Environmental Laboratory, TR-04.1. 127 pp

Livers B, Wohl E (2016) Sources and interpretation of channel complexity in forested subalpine streams of the Southern Rocky Mountains. Water Resour Res 52:3910–3929

Lock MA, Wallace RR, Costerton JW, Ventullo RM, Charton SE (1984) River epilithon: toward a structural-functional model. Oikos 42:10–22

Lowrance R, Todd R, Fail J, Hendrickson O, Leonard R (1984) Riparian forests as nutrient filters in agricultural watersheds. Bioscience 34:374–377

Luck M, Mauemenee N, Whited D et al (2010) Remote sensing analysis of physical complexity of North Pacific Rime rivers to assist wild salmon conservation. Earth Surf Process Landf 35:1330–1343

MacFarlane WA, Wohl E (2003) Influence of step composition on step geometry and flow resistance in step-pool streams of the Washington Cascades. Water Resour Res 39. doi:10.1029/2 001WR001238

MacNally R, Parkinson A, Horrocks G, Conole L, Tzaros C (2001) Relationships between terrestrial vertebrate diversity, abundance and availability of coarse woody debris on south-eastern Australian floodplains. Biol Conserv 99:191–205

Martín-Vide JP, Amarilla M, Zárate FJ (2014) Collapse of the Pilcomayo River. Geomorphology 205:155–163

Massong TM, Montgomery DR (2000) Influence of sediment supply, lithology, and wood debris on the distribution of bedrock and alluvial channels. Geol Soc Am Bull 112:591–599

May CL, Gresswell RE (2003) Processes and rates of sediment and wood accumulation in headwater streams of the Oregon Coast Range, USA. Earth Surf Process Landf 28:409–424

May C, Roering J, Snow K, Griswold K, Greswell R (2017) The waterfall paradox: how knickpoints disconnect hillslope and channel processes, isolating salmonid populations in ideal habitats. Geomorphology 277:228–236

May RM (1977) Thresholds and breakpoints in ecosystems with a multiplicity of stable states. Nature 269:471–477

McClain ME, Boyer EW, Dent CL, Gergel SE, Grimm NB, Groffman PM, Hart SC, Harvey JW, Johnston CA, Mayorga E, McDowell WH, Pinay G (2003) Biogeochemical hot spots and hot moments at the interface of terrestrial and aquatic ecosystems. Ecosystems 6:301–312

Meade RH (2007) Transcontinental moving and storage: the Orinoco and Amazon Rivers transfer the Andes to the Atlantic. In: Gupta A (ed) Large rivers: geomorphology and management. Wiley, Chichester, pp 45–63

Meade RH, Dunne T, Richey JE, Santos UM, Salati E (1985) Storage and remobilization of suspended sediment in the lower Amazon River of Brazil. Science 228:488–490

Meade RH, Moody JA (2010) Causes for the decline of suspended-sediment discharge in the Mississippi River system, 1940–2000. Hydrol Process 24:35–49

Merritt DM (2013) Reciprocal relations between riparian vegetation, fluvial landforms, and channel processes. In: Wohl E (ed) Treatise on fluvial geomorphology. Elsevier, Amsterdam, pp 220–243

Merritts DJ, Vincent KR, Wohl EE (1994) Long river profiles, tectonism, and eustasy: a guide to interpreting fluvial terraces. J Geophys Res 99(B7):14031–14050

Meyer GA, Wells SG, Balling RC, Jull AJT (1992) Response of alluvial systems to fire and climate change in Yellowstone National Park. Nature 357:147–150

Meyer JL, Kaplan LA, Newbold JD, Woltemade CJ, Zedler JB, Beilfuss R, Carpenter Q, Semlitsch R, Watzin MC, Zedler PH (2007) Where rivers are born: the scientific imperative for defending small streams and wetlands. Sierra Club, San Francisco

Meyer JL, Wallace JB (2001) Lost linkages and lotic ecology: rediscovering small streams. In: Press MC, Huntly NJ, Levin S (eds) Ecology: achievement and challenge. Blackwell Science, Orlando, pp 295–317

Miller JR, Friedman JM (2009) Influence of flow variability on floodplain formation and destruction, Little Missouri River, North Dakota. Geol Soc Am Bull 121:752–759

Minckley WL, Rinne JN (1985) Large woody debris in hot-desert streams: an historical review. Desert Plants 7:142–153

Montgomery DR (1999) Process domains and the river continuum. J Am Water Resour Assoc 35:397–410

Montgomery DR, Collins BD, Buffington JM, Abbe TB (2003) Geomorphic effects of wood in rivers. In: Gregory SV, Boyer KL, Gurnell AM (eds) The ecology and management of wood in world rivers. American Fisheries Society Symposium, vol 37, Bethesda, pp 21–47

Muehlbauer JD, Collins SF, Doyle MW, Tockner K (2014) How wide is a stream? Spatial extent of potential 'stream signature' in terrestrial food webs using meta-analysis. Ecology 95:44–55

Nadeau T-L, Rains MC (2007) Hydrological connectivity between headwater streams and downstream waters: how science can inform policy. J Am Water Resour Assoc 43:118–133

Nadler CT, Schumm SA (1981) Metamorphosis of South Platte and Arkansas Rivers, eastern Colorado. Phys Geogr 2:95–115

Naiman RJ, Melillo JM, Hobbie JE (1986) Ecosystem alteration of boreal forest streams by beaver (*Castor canadensis*). Ecology 67:1254–1269

Naiman RJ, Johnston CA, Kelley JC (1988a) Alteration of North American streams by beaver. Bioscience 38:753–762

Naiman RJ, Decamps H, Pastor J, Johnston CA (1988b) The potential importance of boundaries to fluvial ecosystems. J N Am Benthol Soc 7:289–306

Naiman RJ, Decamps H, McClain ME (2005) Riparia: ecology, conservation, and management of streamside communities. Elsevier, Amsterdam

Nakano S, Murakami M (2001) Reciprocal subsidies: dynamic interdependence between terrestrial and aquatic food webs. Proc Natl Acad Sci U S A 98:166–170

Nanson GC (1986) Episodes of vertical accretion and catastrophic stripping: a model of disequilibrium flood-plain development. Geol Soc Am Bull 97:1467–1475

Nanson GC, Croke JC (1992) A genetic classification of floodplains. Geomorphology 4:459–486

Newbold JD, Elwood JW, O'Neill RV, Van Winkle W (1981) Measuring nutrient spiralling in streams. Can J Fish Aquat Sci 38:860–863

Nihlgard BJ, Swank WT, Mitchell MJ (1994) Biological processes and catchment studies. In: Moldan B, Cerny J (eds) Biogeochemistry of small catchments: a tool for environmental research. Wiley, Chichester, pp 133–161

Nilsson C, Lepori F, Malmqvist B, Törnlund E, Hjerdt N, Helfield JM, Palm D, Östergren J, Jansson R, Brännäs E, Lundqvist H (2005) Forecasting environmental responses to restoration of rivers used as log floatways: an interdisciplinary challenge. Ecosystems 8:779–800

O'Connor JE, Jones MA, Haluska TL (2003) Flood plain and channel dynamics of the Quinault and Queets Rivers, Washington, USA. Geomorphology 51:31–59

Omernik JM (2004) Perspectives on the nature and definition of ecological regions. Environ Manag 34:27–38

Osterkamp WR, Hedman ER (1977) Variation of width and discharge for natural high-gradient stream channels. Water Resour Res 13:256–258

Owens PN, Walling DE (2002) Changes in sediment sources and floodplain deposition rates in the catchment of the River Tweed, Scotland, over the last 100 years: the impact of climate and land use change. Earth Surf Process Landf 27:403–423

Palmer MA, Menninger HL, Bernhardt E (2010) River restoration, habitat heterogeneity and biodiversity: a failure of theory or practice? Freshw Biol 55:205–222

Parrish JD, Braun DP et al (2003) Are we conserving what we say we are? Measuring ecological integrity within protected areas. Bioscience 53:851–860

Pazzaglia FJ (2013) Fluvial terraces. In: Wohl E (ed) Treatise on fluvial geomorphology. Treatise on geomorphology, vol 9. Academic, San Diego, pp 379–412

Peipoch M, Brauns M, Hauer RF, Weitere M, Valett HM (2015) Ecological simplification: human influences on riverscape complexity. Bioscience 65:1057–1065

Pettit NE, Naiman RJ (2006) Flood-deposited wood creates regeneration niches for riparian vegetation on a semiarid South African river. J Veg Sci 17:615–624

Phillips JD (2007) Perfection and complexity in the lower Brazos River. Geomorphology 91:364–377

Phillips JD (2014) State transitions in geomorphic responses to environmental change. Geomorphology 204:208–216

Poff NL, Allan JD, Bain MB, Karr JR, Prestegaard KL, Richter BD, Sparks RE, Stromberg JC (1997) The natural flow regime: a paradigm for river conservation and restoration. Bioscience 47:769–784

Poff NL, Richter BD, Arthington AH, Bunn SE, Naiman RJ et al (2010) The ecological limits of hydrologic alteration (ELOHA): a new framework for developing regional environmental flow standards. Freshw Biol 55:147–170

Poff NL, Matthews JH (2013) Environmental flows in the Anthropocene: past progress and future prospects. Curr Opin Environ Sustain 5:667–675

Pollen N, Simon A (2005) Estimating the mechanical effects of riparian vegetation on stream bank stability using a fiber bundle model. Water Resour Res 41:W07025. doi:10.1029/2004WR003801

Pollock MM, Beechie TJ, Jordan CE (2007) Geomorphic changes upstream of beaver dams in Bridge Creek, an incised stream channel in the interior Columbia River basin, eastern Oregon. Earth Surf Process Landf 32:1174–1185

Pollock MM, Beechie TJ, Wheaton JM, Jordan CE, Bouwes N, Weber N, Volk C (2014) Using beaver dams to restore incised stream ecosystems. Bioscience 64:279–290

Polvi LE, Wohl E (2012) The beaver meadow complex revisited—the role of beavers in post-glacial floodplain development. Earth Surf Process Landf 37:332–346

Polvi LE, Wohl E (2013) Biotic drivers of stream planform: implications for understanding the past and restoring the future. Bioscience 63:439–452

Polvi LE, Nilsson C, Hasselquist EM (2014) Potential and actual geomorphic complexity of restored headwater streams in northern Sweden. Geomorphology 210:98–118

Poole GC (2002) Fluvial landscape ecology: addressing uniqueness within the river discontinuum. Freshw Biol 47:641–660

Pringle CM (2001) Hydrologic connectivity and the management of biological reserves: a global perspective. Ecol Appl 11:981–998

Pringle CM, Naiman RJ, Bretschko G, Karr JR, Oswood MW, Webster JR, Welcomme RL, Winterbourn MJ (1988) Patch dynamics in lotic systems: the stream as a mosaic. J N Am Benthol Soc 7:503–524

Reid HE, Brierley GJ (2015) Assessing geomorphic sensitivity in relation to river capacity for adjustment. Geomorphology 251:108–121

Richmond AD, Fausch KD (1995) Characteristics and function of large woody debris in subalpine Rocky Mountain streams in northern Colorado. Can J Fish Aquat Sci 52:1789–1802

Richter BD, Baumgartner J, Powell J, Braun D (1996) A method for assessing hydrologic alteration within ecosystems. Conserv Biol 10:1163–1174

Ripple WJ, Beschta RL (2004) Wolves and the ecology of fear: can predation risk structure ecosystems? Bioscience 54:755–766

Sawyer AH, Cardenas MB, Buttles J (2011) Hyporheic exchange due to channel-spanning logs. Water Resour Res 47:W08502. doi:10.1029/2011WRR010484

Scanlon BR, Jolly I, Sophocleous M, Zhang L (2007) Global impacts of conversions from natural to agricultural ecosystems on water resources: quantity versus quality. Water Resour Res 43:W03437. doi:10.1029/2006WR005486

Scheffer M, Carpenter SR (2003) Catastrophic regime shifts in ecosystems: linking theory to observation. Trends Ecol Evol 18:648–656

Schmidt JC, Wilcock PR (2008) Metrics for assessing the downstream effects of dams. Water Resour Res 44:W04404. doi:10.1029/2006WR005092

Schröder A, Persson L, De Roos AM (2005) Direct experimental evidence for alternative stable states: a review. Oikos 110:3–19

Schumm SA (1969) River metamorphosis. J Hydraul Div ASCE 95:255–273

Schumm SA (1973) Geomorphic thresholds and complex response of drainage systems. In: Morisawa M (ed) Fluvial geomorphology. SUNY Binghamton, New York, pp 299–310

Schumm SA (1979) Geomorphic thresholds: the concept and its applications. Trans Inst Br Geogr 4:485–515

Schumm SA, Hadley RF (1957) Arroyos and the semiarid cycle of erosion. Am J Sci 255:161–174

Scott ML, Skagen SK, Merigliano MF (2003) Relating geomorphic change and grazing to avian communities in riparian forests. Conserv Biol 17:284–296

Shields FD, Simon A, Steffen LJ (2000) Reservoir effects on downstream river channel migration. Environ Conserv 27:54–66

Sklar LS, Dietrich WE (2004) A mechanistic model for river incision into bedrock by saltating bed load. Water Resour Res 40:W06301. doi:10.1029/2003WR002496

Small MJ, Doyle MW, Fuller RL et al (2008) Hydrologic versus geomorphic limitation on CPOM storage in stream ecosystems. Freshw Biol 53:1618–1631

Stanford JA, Ward JV (1988) The hyporheic habitat of river ecosystems. Nature 335:64–66

Stock JD (2013) Waters divided: a history of alluvial fan research and a view of its future. In: Wohl E (ed) Treatise on fluvial geomorphology. Treatise on geomorphology, vol 9. Academic, San Diego, pp 413–458

Suding KN, Gross KL, Houseman GR (2004) Alternative states and positive feedbacks in restoration ecology. Trends Ecol Evol 19:46–53

Thorp JH, Thoms MC, Delong MD (2006) The riverine ecosystem synthesis: biocomplexity in river networks across space and time. River Res Appl 22:123–147

Tockner K, Malard F, Ward JV (2000) An extension of the flood-pulse concept. Hydrol Process 14:2861–2883

Tonina D, Buffington JM (2009) Hyporheic exchange in mountain rivers I: mechanics and environmental effects. Geogr Compass 3:1063–1086

Trimble SW (2013) Historical agriculture and soil erosion in the Upper Mississippi Valley Hill Country. CRC Press, Boca Raton, p 242

Trimble SW, Mendel AC (1995) The cow as a geomorphic agent—a critical review. Geomorphology 13:233–253

Triska FJ (1984) Role of wood debris in modifying channel geomorphology and riparian areas of a large lowland river under pristine conditions: a historical case study. Verh Int Ver Limnol 22:1876–1892

U.S. Army Corps of Engineers (USACE) (2012) 2012 Nationwide permits, conditions, district engineer's decision, further information, and definitions. http://www.usace.army.mil/Portals/2/docs/civilworks/nwp/2012/NWP2012_corrections_21-sep-2012.pdf

Van Breemen N, Boyer EW, Goodale CL, Jaworski NA, Paustian K, Seitzinger SP, Lajtha K et al (2002) Where did all the nitrogen go? Fate of nitrogen inputs to large watersheds in the northeastern U.S.A. Biogeochemistry 57/58:267–293

Vannote RL, Minshall GW, Cummins KW, Sedell JR, Cushing CE (1980) The river continuum concept. Can J Fish Aquat Sci 37:130–137

Wainwright J, Turnbull L, Ibrahim TG, Lexartza-Artza I, Thomas SF, Brazier RE (2011) Linking environmental regimes, space, and time: interpretations of structural and functional connectivity. Geomorphology 126:387–404

Ward JV (1989) The four-dimensional nature of lotic ecosystems. J N Am Benthol Soc 8:2–8

Ward JV (1992) A mountain river. In: Calow P, Petts GE (eds) The rivers handbook. Blackwell Science, Oxford, pp 493–510

Ward JV, Stanford JA (1983) The serial discontinuity concept of lotic ecosystems. In: Fontaine TD, Bartell SM (eds) Dynamics of lotic ecosystems. Ann Arbor Science, Ann Arbor, pp 29–42

Ward JV, Stanford JA (1995) The serial discontinuity concept: extending the model to floodplain rivers. Regul Rivers Res Manag 10:159–168

Ward JV, Tockner K, Schiemer F (1999) Biodiversity of floodplain river ecosystems: ecotones and connectivity. Regul Rivers Res Manag 15:125–139

Webster JR, Patten BC (1979) Effects of watershed perturbation on stream potassium and calcium dynamics. Ecol Monogr 49:51–72

Webster JR, Waide JB, Pattern BC (1975) Nutrient recycling and the stability of ecosystems. In: Howell FG et al (eds) Mineral cycling in southeastern ecosystems. ERDA (CONF-740513), pp 1–27

Werner BT, McNamara DE (2007) Dynamics of coupled human-landscape systems. Geomorphology 91:393–407

Westbrook CJ, Cooper DJ, Baker BW (2006) Beaver dams and overbank floods influence groundwater-surface water interactions of a Rocky Mountain riparian area. Water Resour Res 42:206404

Westbrook CJ, Cooper DJ, Butler DR (2013) Beaver hydrology and geomorphology. In: Butler DR, Hupp CR (eds) Ecogeomorphology. Treatise on geomorphology, vol 12. Academic, San Diego, pp 293–306

White PS, Pickett STA (1985) Natural disturbance and patch dynamics: an introduction. In: Pickett STA, White PS (eds) The ecology of natural disturbance and patch dynamics. Academic, New York, pp 3–13

White PS, Walker JL (1997) Approximating Nature's variation: selecting and using reference information in restoration ecology. Restor Ecol 5:338–349

Wilcock PR, Kondolf GM, Matthews WVG, Barta AF (1996) Specification of sediment maintenance flows for a large gravel-bed river. Water Resour Res 32:3911–3921

Williams GP (1978a) Bank-full discharge in rivers. Water Resour Res 14:1141–1154

Williams GP (1978b) The case of the shrinking channels—the North Platte and the Platte Rivers in Nebraska. U.S. Geological Survey Circular 781, Arlington, VA

Wohl E (2002) Modeled paleoflood hydraulics as a tool for interpreting bedrock channel morphology. In: House PK, Webb RH, Baker VR, Levish DR (eds) Ancient Floods, modern hazards: principles and applications of paleoflood hydrology. American Geophysical Union Press, Washington, DC, pp 345–358

Wohl E (2010) Analysing a natural system. In: Clifford N, French S, Valentine G (eds) Key methods in geography, 2nd edn. Sage, London, pp 253–273

Wohl E (2011) Threshold-induced complex behavior of wood in mountain streams. Geology 39:587–590

Wohl E (2013) The complexity of the real world in the context of the field tradition in geomorphology. Geomorphology 200:50–58

Wohl E (2014a) A legacy of absence: wood removal in U.S. rivers. Prog Phys Geogr 38:637–663

Wohl E (2014b) Rivers in the landscape: science and management. Wiley Blackwell, Chichester

Wohl E (2016) Spatial heterogeneity as a component of river geomorphic complexity. Prog Phys Geogr 40:598–615

Wohl E, Scott DN (2017) Wood and sediment storage and dynamics in river corridors. Earth Surf Process Landf 42:5–23

Wohl E, Dwire K, Sutfin N, Polvi L, Bazon R (2012) Mechanisms of carbon storage in mountainous headwater rivers. Nat Commun 3:1263. doi:10.1038/ncomms2274

Wohl E, Bledsoe BP, Jacobson RB, Poff NL, Rathbun SL, Walters DM, Wilcox AC (2015) The natural sediment regime in rivers: broadening the foundation for ecosystem management. Bioscience 65:358–371

Wohl E, Mersel MK, Allen AO, Fritz KM, Kichefski SL, Lichvar RW, Nadeau TL, Topping BJ, Trier PH, Vanderbilt FB (2016) Synthesizing the scientific foundation for ordinary high water mark delineation in fluvial systems. Cold Regions Research and Engineering Laboratory, U.S. Army Corps of Engineers, ERDC/CRREL SR-16-5, Washington, 198 pp

Wohl EE (1992) Bedrock benches and boulder bars: floods in the Burdekin Gorge of Australia. Geol Soc Am Bull 104:770–778

Wolf EC, Cooper DJ, Hobbs NT (2007) Hydrologic regime and herbivory stabilize an alternative state in Yellowstone National Park. Ecol Appl 17:1572–1587

Wondzell SM, Swanson FJ (1996) Seasonal and storm dynamics of the hyporheic zone of a fourth-order mountain stream. 1. Hydrologic processes. J N Am Benthol Soc 15:3–19

Wondzell SM, LaNier J, Haggerty R, Woodsmith RD, Edwards RT (2009) Changes in hyporheic exchange flow following experimental wood removal in a small, low-gradient stream. Water Resour Res 45:W05406. doi:10.1029/2008WR007214

Woods SW, MacDonald LH, Westbrook CJ (2006) Hydrologic interactions between an alluvial fan and a slope wetland in the central Rocky Mountains, USA. Wetlands 26:230–243

World Wildlife Fund (WWF) (2013) Freshwater ecoregions of the world. http://www.feow.org/index.php

Yochum SE, Bledsoe BP, David GCL, Wohl E (2012) Velocity prediction in high-gradient channels. J Hydrol 424–425:84–98

Chapter 3
Human Alterations of Rivers

People have been altering the environment since prehistory. Archeological records and the stratigraphy of valley bottoms suggest that prehistoric alteration of native upland vegetation for grazing and crops resulted in enhanced sediment yields to river corridors, as well as associated changes in channel dimensions and planform and channel-floodplain connectivity (e.g., Mei-e and Xianmo 1994; Stinchcomb et al. 2011). The magnitude of these changes varied through time and space. Examples of intensive vegetation clearing and valley-bottom aggradation come from diverse regions. In the southeastern United States, nineteenth century row-crop agriculture led to floodplain deposition of more than a meter of sediment (Jackson et al. 2005). In southeastern Australia, nineteenth century land clearance caused channel and floodplain aggradation (Brooks and Brierley 1997). In southern Poland, land clearance for agriculture during the late seventeenth to nineteenth centuries triggered increased sediment yields that changed meandering rivers into braided channels (Latocha and Migoń 2006). Conversely, twentieth century declining resource use, regrowth of upland vegetation, and erosion of channels occurred in mountainous regions of western and central Europe (Latocha and Migoń 2006).

Understanding the history and characteristics of environmental alteration are important for effective river management because even activities undertaken more than a century ago and physically outside of the river corridor can continue to influence river form and process in ways that are not always readily apparent. This chapter reviews the plethora of human activities that have directly and indirectly affected river ecosystems.

In the context of river networks and river corridors, human environmental alteration can be distinguished as indirect or direct. Indirect alterations are those within the drainage basin but outside of the river corridor. Changes in land cover or climate, for example, can alter water and sediment yields to river corridors, as well as nutrient inputs and riverine biota. Direct alterations are those within the river network. Direct alterations affect river corridor configuration and connectivity, as well as water and sediment regimes. Explicitly distinguishing these categories of altera-

© The Author(s) 2018
E. Wohl, *Sustaining River Ecosystems and Water Resources*, SpringerBriefs in
Environmental Science, DOI 10.1007/978-3-319-65124-8_3

Table 3.1 Indirect alterations of river corridors and the most commonly observed, general effects in terms of altered yields of water, sediment, nutrients, and large wood (LW) from uplands to river corridors

Alteration	Effects on yields to river corridors	Sample references
Climate change	Variable, but can include altered water, sediment, nutrients and LW	Mirza et al. (2003), Stewart et al. (2005), Rood et al. (2008), Schmocker-Fackel and Naef (2010), Goode et al. (2012) and Nilsson et al. (2015)
Change in land cover Deforestation Afforestation Upland grazing Upland cropping Changes in fire	+ Sediment, + water, – LW – Sediment, – water, + LW + Sediment, + water + Sediment, + water, + nutrients Variable, but alters water, sediment, nutrients, and LW	Nik (1988), Luce and Black (1999) and Fransen et al. (2001) Keesstra et al. (2005) Trimble and Mendel (1995) Knox (1977) and de Boer (1997) MacDonald et al. (2000), May and Gresswell (2003) and Pierce et al. (2004)
Change in topography Mining Road construction	+ Sediment + Sediment	Rooney et al. (2012) and Wickham et al. (2013) Larsen and Parks (1997)
Introduced species	Variable, depending on extent and type of species, but include altered water, sediment, nutrient, and LW	Baillie and Davies (2002) and Ulloa et al. (2011)
Water diversions between channels or catchments	– Water yield in source basin, + water yield in receiving basin	Ryan (1997), Baker et al. (2011), Wohl and Dust (2012), and Gabbud and Lane (2015)
Urbanization	– Sediment, + water, + nutrients	Wolman (1967), Roberts (1989) and Bledsoe and Watson (2001)

tions provides a useful reminder that activities distant from a channel, but within the drainage basin, can substantially alter river process and form. Indirect and direct alterations are commonly interlinked, such as in scenarios where changes in upland vegetation increase sediment yield to channels, resulting in channelization to increase downstream conveyance (Sutter 2015). Tables 3.1 and 3.2 lists the diverse forms of indirect and direct alteration.

3.1   Indirect Alterations of River Networks and River Corridors

Recognition of indirect alterations of river networks and river corridors is particularly important for management of rivers as ecosystems because the effect of these alterations is likely to be less obvious than the effect of activities

Table 3.2 Direct alterations of river corridors

Alteration	Effects	Sample references

Channel engineering
  Dredging
  Bank stabilization
  Channelization
  Wood removal
	Decreased hydraulic resistance; increased downstream conveyance and flow velocity; decreased complexity, lateral and vertical connectivity, and retention; bed coarsening and channel incision	Scarnecchia (1988), Rhoads (1990), Kesel and Yodis (1992), Wyzga (2001) and Hohensinner et al. (2004)

Placer and aggregate mining	Altered channel geometry and stability; increased sediment transport	Knighton (1989), Hilmes and Wohl (1995) and Lawrence and Davies (2014)
Beaver trapping	Decreased hydraulic resistance; increased downstream conveyance and flow velocity; decreased complexity, lateral and vertical connectivity, and retention	Pollock et al. (2003, 2007, 2014), Green and Westbrook (2009) and Polvi and Wohl (2012, 2014)
Log floating	Simplified, enlarged channel cross sectional area; decreased hydraulic resistance; increased downstream conveyance and flow velocity; decreased complexity, lateral and vertical connectivity, and retention	Young et al. (1994), Törnlund and Östlund (2002), Comiti (2012) and Haidvogl et al. (2015)
Floodplain drainage	Decreased complexity and retention, particularly with respect to nutrients	Spaling and Smit (1995) and Holden (2006)
Levee construction	Increased downstream conveyance and flow velocity; decreased complexity, lateral connectivity, and retention; bed coarsening and channel incision	Frings et al. (2009) and Czech et al. (2015)
Floodplain deforestation	Decreased complexity and retention; increased bank erosion	Iwata et al. (2003)
Riparian grazing	Decreased riparian vegetation, increased runoff and bank erosion, wider channel cross section	Trimble and Mendel (1995)
Flow regulation	Altered hydrograph, transport capacity, water chemistry, connectivity	Ligon et al. (1995), Magilligan and Nislow (2001), Poff et al. (2007) and Kondolf et al. (2014)
Introduced species	Variable, depending on extent and type of species, but include altered hydraulic and bank resistance, and altered sediment transport, and nutrient and LW availability	Graf (1978) and Ulloa et al. (2011)
Commercial extinction of species (fish, mussels, etc)	Changes in river trophic cascades, primary production, and suspended sediment	Strayer et al. (1999, 2004)

undertaken within the river corridor. Indirect alterations such as changes in upland vegetation, however, can strongly influence inputs of water, sediment, and nutrients to the river corridor and thus strongly influence river ecosystem function. This section reviews the different types of indirect alterations and their effects on river ecosystems.

The basic categories of indirect alterations are climate change; changes in land cover; changes in topography; introduced upland species; water diversions from other catchments; and urbanization (Table 3.1), which commonly includes changes in land cover, topography, and species. The details of the magnitude and duration of changes in water, sediment, nutrients, and large wood yields to river corridors vary between specific sites, but the general direction (increase or decrease) of changes in response to indirect alterations is consistent for many of these categories. Deforestation and agriculture, for example, typically increase sediment yields to river corridors, whereas urbanization increases water yields. Climate change is an exception in that its effects are difficult to generalize between regions. Although global average temperatures are increasing, the effects of this increase on specific meteorological parameters such as mean annual precipitation or precipitation intensity and seasonality vary significantly between different watersheds around the world, as illustrated by two specific examples (Fig. 3.1).

A confounding effect of climate change is that other human alterations of landscapes and river networks can limit the ability of species to migrate to or colonize more suitable habitats if climate change renders a particular river segment unsuitable for the species (Palmer et al. 2009). Endangered bull trout (*Salvelinus confluentus*) in the Northern Rockies of the United States, for example, require relatively cold water temperatures that typically occur at higher elevations within mountainous river networks. As warming air temperatures reduce the abundance of thermally suitable portions of river networks, human alterations such as flow regulation, dams, and diversions that limit longitudinal connectivity or change water temperature can prevent bull trout from migrating between suitable habitats. This causes isolated populations to become genetically inbred and makes these populations more susceptible to local extinction associated with a severe natural or human-induced disturbance (Davies 2010; Wenger et al. 2011; Isaak et al. 2012). Similar scenarios have been described for freshwater fauna in global biodiversity hot spots such as southwestern Australia (Davies 2010). Restoration options that focus on restoring connectivity and natural flow regimes are more likely to ameliorate climate-related changes in flow and temperature and to increase population resilience than are restoration activities focused on within-channel rehabilitation (Beechie et al. 2013). Rehabilitation within a channel can increase habitat abundance and diversity, but this may be of limited use if species cannot migrate to other portions of a river network as climate change makes a river segment unsuitable for the species.

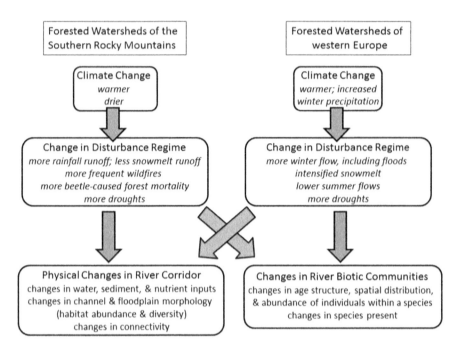

Fig. 3.1 Schematic illustration of cascade of climate change effects for watersheds in the U.S. Southern Rocky Mountains (e.g., the Upper South Plate River basin) and watersheds in western Europe (e.g., the Rhine River basin)

3.1.1 Changes in Land Cover

Changes in land cover within a drainage basin but outside of the river network include removal of forest vegetation; natural or human-induced growth of forest cover (e.g., commercial tree plantations); grazing; cropping; and altered fire regime. Each of these can change the relations among precipitation inputs, infiltration, runoff, and evapotranspiration, and thus change water yields—volume, mechanism of delivery (diffuse surface flow, diffuse or conduit subsurface flow), speed, episodicity—to river corridors (Fig. 3.2). Liu et al. (2008) illustrate these effects across China, where rapid change in land cover during the twentieth century has changed evapotranspiration and water yield. In general, deforestation in China has increased evapotranspiration and decreased water yield, whereas urbanization has decreased evapotranspiration and increased water yield.

Changes in land cover also affect the interception of precipitation, ground cover, and soil compaction. Through these alterations, changes in land cover can substantially change sediment yields with respect to volume, grain-size distribution, and episodicity of sediment inputs (Fig. 3.3). The magnitude and effects of the changes in

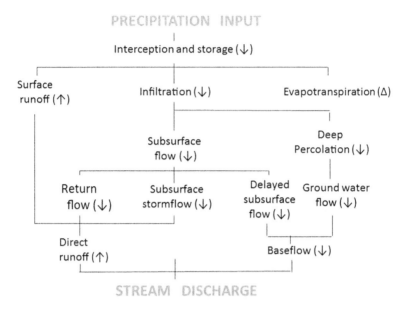

Fig. 3.2 Flow chart illustrating the most common changes in how precipitation enters rivers after changes in land cover. *Upward arrow* indicates increase; *downward arrow* indicates decrease; *triangle* indicates change (evapotranspiration can either increase or decrease following changes in land cover, depending on what replaces the native plant community)

sediment yield depend on the characteristics of the river corridor. Summarizing multiple studies conducted in mountainous areas of New Zealand, for example, Liébault et al. (2005) describe how the conversion of forest to pasture between AD 1840 and 1940 resulted in an order of magnitude increase in sediment yield. The sediment was rapidly transported through the steep, narrow mountainous valleys, however, in contrast to river networks in other regions of the world that have experienced channel aggradation and widening following deforestation (Liébault et al. 2005).

Changes in land cover alter nutrient yields to the river corridor by altering the primary productivity of upland vegetation; the nutrient content of soils; and the transport of nutrients moving downslope with water and sediment. Cropping, in particular, typically increases nutrient yields because of excess fertilizer transported from crop fields in dissolved form during runoff (Vitousek et al. 2009), but any change in land cover can alter nutrient yields.

In regions that are forested under natural conditions, changes in land cover can also significantly alter the recruitment of large wood to river corridors by increasing or decreasing forest cover and altering slope stability. Deforestation in regions of steep terrain, for example, is likely to increase the frequency and size of landslides and debris flows (Imaizumi et al. 2008). These forms of slope instability can recruit large wood to river corridors (May and Gresswell 2003), so one indirect effect of deforestation can be a short-term increase in wood inputs from unstable hillslopes that have not been completely deforested.

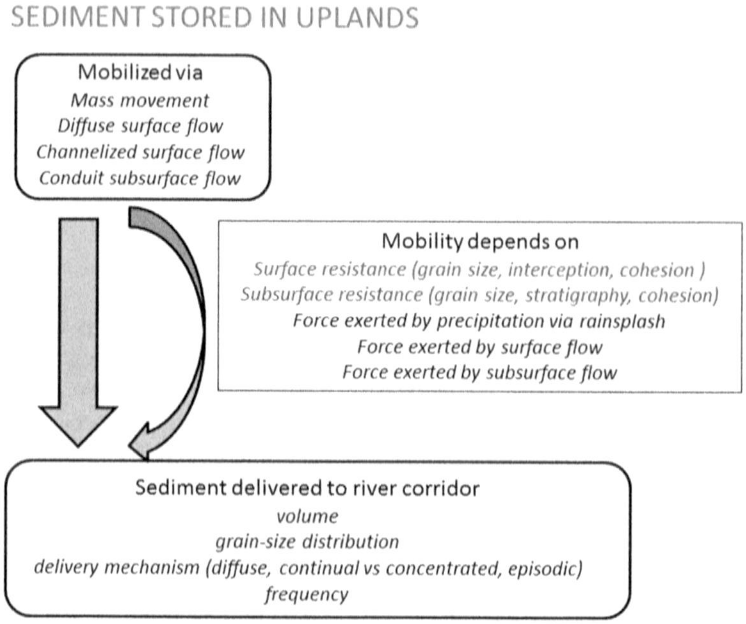

SEDIMENT STORED IN UPLANDS

Mobilized via

Mass movement
Diffuse surface flow
Channelized surface flow
Conduit subsurface flow

Mobility depends on

Surface resistance (grain size, interception, cohesion)
Subsurface resistance (grain size, stratigraphy, cohesion)
Force exerted by precipitation via rainsplash
Force exerted by surface flow
Force exerted by subsurface flow

Sediment delivered to river corridor
volume
grain-size distribution
delivery mechanism (diffuse, continual vs concentrated, episodic)
frequency

Fig. 3.3 Flow chart illustrating the controls on sediment inputs from uplands to the river corridor. Variables affected by changes in land cover are in *brown* text. This illustrates how changes in land cover alter surface and subsurface resistance of sediment to downslope movement

3.1.2 Changes in Upland Topography

Changes in upland topography can occur under any form of land use, such as agricultural terracing for crops, but the most substantial changes are likely to occur in connection with mining and construction of transportation corridors. Mining includes underground removal of material, open-pit mines, and mountaintop removal. The primary physical effect on river networks of underground mining is an increase in sediment yield from tailings piles (Macklin et al. 2003). Chemical effects include leaching of toxic contaminants into the river corridor (Stoughton and Marcus 2000; Hren et al. 2001). Open-pit mines can significantly alter hillslope surface and subsurface flow paths and thus alter water yields, but the most consistent effects are increased sediment yields from erosion of tailings and waste rock, and chemical contamination (Dudko and Adriano 1995; Schreck 1998). Mountaintop removal, as practiced in the United States in association with coal mining, completely obliterates large portions of hillslopes and fills adjacent valleys with waste rock. Consequently, this form of mining substantially alters water, sediment, nutrient, and large wood yields to river corridors (Palmer et al. 2010).

Construction of major transportation corridors such as railroads and large roads can involve moving large amounts of sediment and disrupting native land cover, especially in high-relief terrain. These changes can result in continuing hillslope instability and associated changes in water, sediment, and nutrient yields to river corridors (e.g., Larsen and Parks 1997).

3.1.3 Introduced Upland Species

Introduced upland species, typically plant species, can change relations among precipitation inputs, infiltration, runoff, and evapotranspiration, and thus change water, sediment, and nutrient yields and, in the case of woody plants, large wood yields to river corridors. Examples come from regions with introduced tree species, such as Chile and New Zealand, where forested streams with introduced tree species have less large wood and/or less geomorphically effective large wood (e.g., Evans et al. 1993; Baillie and Davies 2002; Ulloa et al. 2011). Introduced insects such as the woolly adelgid (*Adelges tsugae*) or emerald ash borer (*Agrilus planipennis*) in North America, both of which kill large numbers of trees, can substantially alter evapotranspiration and water fluxes, with resulting changes in stream flow. Widespread tree die-off from insect infestation also alters inputs of large wood and nutrients to river networks (e.g., Ford and Vose 2007).

3.1.4 Water Diversions

People have transferred water from one drainage basin to another for centuries in some parts of the world (Liu and Zheng 2002). Long-distance water transfers are now becoming increasingly widespread as the technology to facilitate such transfers, and the economic incentive to support them, develop (Wohl 2011). The most obvious effect of a water transfer is to decrease flow in the source basin and increase flow in the receiving basin, with a cascade of secondary effects on transport of sediment and nutrients; complexity; and connectivity in source and receiving basins (Fig. 3.4). The details of these effects depend on what proportion of total flow is diverted and how that flow is transferred to the receiving basin (e.g., whether the flow is transferred throughout the year or for a limited period) (Ryan 1997; Wohl and Dust 2012).

3.1.5 Urbanization

Urbanization commonly includes changes in land cover, topography, and species within the urban area. Among the most extensively documented effects of urbanization are a substantial increase in the volume of water delivered to river

Fig. 3.4   2016 view of the Owens River downstream from the Los Angeles Aqueduct, illustrating effects of flow diversion into the aqueduct. Extensive *gray* area around the relatively narrow wetted portion of the channel indicates extent of the river corridor prior to flow diversion via the aqueduct. *Yellow arrow* indicates flow direction. (Photograph courtesy of Google Earth)

corridors as a result of increased impervious area, and an increase in the speed with which precipitation inputs reach the river corridor as a result of enhanced overland flow and concentration in storm drains (DeWalle et al. 2000; Ogden et al. 2011). Increased impervious area commonly also results in decreased sediment yields. Reduction in woody vegetation or active removal of large wood results in reduced large wood yields. Household and industrial uses of nitrogen and phosphorus substantially increase nutrient yields to river corridors, as well as yields of an extremely broad array of synthetic chemicals, heavy metals, pathogens, and other toxic materials (Pitt et al. 1995; Mallin et al. 2008). The net effects of urbanization on the river ecosystem partly depend on what proportion of the drainage basin is urbanized.

   Urbanization can also directly affect a river ecosystem if the river corridor passes through the urban area. In this case, alterations such as channel engineering, stream burial, storm drains, and storm sewers can enhance longitudinal connectivity of water, but limit or sever longitudinal connectivity for aquatic organisms (Blakely et al. 2006).

3.2   Direct Alterations of River Networks and River Corridors

Direct alterations are those that occur within the river corridor. These alterations can affect the form of the channel or floodplain, as in channel engineering, mining within the river corridor, log floating, and floodplain drainage. Direction alterations can also change the fluxes of materials and the movement of organisms within the corridor, as in levees and flow regulation. Finally, direct alterations can change the presence of particular species within the corridor, as in the case of beaver removal, introduced species, and commercial extinction of species. Recognition of direct alterations is important for river management because these activities may have occurred so long ago that, although their effects persist, they are no longer immediately apparent, as in the case of abandoned, historical mill dams discussed in Sect. 3.2.2.

Each of the types of direct alteration of river corridors listed above commonly affects the other types. Channelization, for example, simplifies and homogenizes the channel. This increases downstream fluxes and limits lateral and vertical fluxes, which may decrease habitat availability and stability and thus alter aquatic and riparian biomass and biodiversity (Schoof 1980; Fig. 3.5). Introduced riparian vegetation may substantially increase hydraulic and erosional resistance of stream banks and reduce stream flow because of increased water uptake and transpiration (e.g., tamarisk (*Tamarix* spp.) in the southwestern United States (Graf 1978) and willow (*Salix* spp.) in Australia (Schulze and Walker 1997; Cremer 2003)). Introduced riparian vegetation can also change nutrient cycling and habitat available for riverine species (Greenwood et al. 2004). As with the effects of indirect alterations, direct alterations of the river corridor seldom have just one effect. Instead, the initial alteration triggers a cascade of interrelated changes that can make prediction and management of the river corridor very challenging (Fig. 3.6).

3.2.1   Changes in Channel and Floodplain Form

This section briefly reviews channel engineering, mining within the river corridor, log floating, and floodplain drainage. Although other types of direct changes in channel and floodplain form exist, the four categories of change discussed here are among the most common and widespread.

Among the forms of direct alterations, channel engineering has a particularly long history. Channel engineering refers to activities designed to simplify and homogenize the channel form and to 'harden' the channel boundaries in order to make them more resistant to erosion and to limit lateral or vertical channel changes. Types of channel engineering include dredging of the channel bed, usually to enhance navigation or downstream conveyance of flood waters or waste products. Another form of channel engineering is stabilization of the channel banks, usually

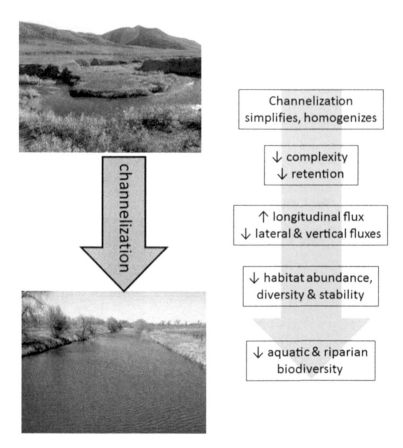

Fig. 3.5 Schematic illustration of cascade of effects on channel form, complexity, connectivity, retention, and biodiversity as a result of channelization. Upper photo shows a diverse, natural river in the Brooks Range of Alaska, USA. Lower photo shows a portion of the Cache la Poudre River, Colorado, USA that was channelized during the 1960s; photograph was taken 50 years after channelization

to limit property loss or hazards associated with channel movement. Channelization is yet another form of channel engineering, which involves widening, deepening, and straightening the channel, but also blocking the entrance to secondary channels. Finally, removal of instream and bank large wood is a form of channel engineering. These activities have most commonly been undertaken to limit overbank flooding and the duration of flooding; to enhance navigation; and to limit channel erosion of the floodplain (DuPont 1925; Gillette 1972; Landwehr and Rhoads 2003; Hohensinner et al. 2004). Channelization has also sometimes been conducted to enhance fish passage or to make the channel more esthetically attractive.

Records of activities such as bank stabilization and large wood removal go back centuries in Eurasia (Comiti 2012) and North America (Wohl 2014). In the United States, for example, the national government removed at least 1.5 million pieces of

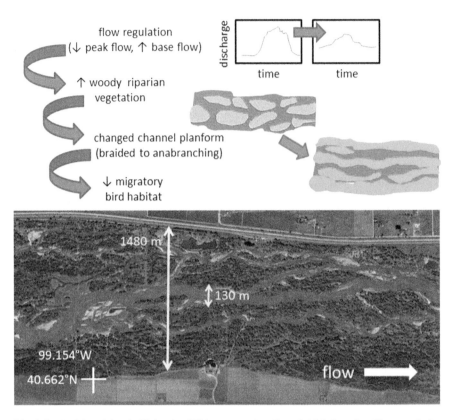

Fig. 3.6 The Platte River in Nebraska, USA as example of how initial alteration (flow regulation that decreased peak flow and increased base flow) created a cascade of changes in the river corridor that are now difficult to reverse with management. In upper panel, orange patches within channel in planform drawing indicate unvegetated sand bars and green patches indicate vegetated islands and channel margins. Aerial photograph shows the Platte River in 2015 near Kearney, Nebraska. *Long yellow arrow* indicates lateral extent of formerly active braided channel, and *short yellow arrow* indicates the width of one of the contemporary channel anabranches. (Photograph courtesy of Google Earth)

large wood from major rivers between 1867 and 1912 (Harmon et al. 1986; Wohl 2014). Physical and ecological effects of the lack of large wood persist (Collins et al. 2002), but contemporary societies may not be aware that large wood quantities of wood were once present and were removed.

The net effect of channel engineering is to concentrate flow in a channel that more efficiently conveys water downstream. This equates to reductions in channel complexity, habitat abundance and diversity, retention, lateral and vertical connectivity, and channel stability, all of which lead to reduced biomass and biodiversity of aquatic and riparian organisms (Ward 1998; Habersack and Piégay 2008). Channel engineering has been so widespread and intensive that the multibillion dollar river restoration industry now present in Europe and North America is largely designed to mitigate the many negative effects of channelization, as discussed in Chap. 4.

Fig. 3.7 2016 aerial view of aggregate mining within the river corridor of the Rio Maipo near Santiago, Chile. *Yellow arrow* indicates flow direction. Inset photo at upper right is an upstream view and photo at lower right is a downstream view from ground level, both taken in 2012. (Aerial photograph courtesy of Google Earth)

Placer and aggregate mining refer to removal of sediment from the river corridor. Placer metals are precious metals—usually gold—disseminated through alluvial sediment in the channel and floodplain. Aggregate is sand and gravel used in various forms of construction. In each case, mining involves displacing or removing massive quantities of channel and floodplain sediment from the river corridor. This displacement and removal cause changes in valley-bottom topography and increase sediment supply and mobility. This in turn decreases habitat abundance and stability, as well as biomass and biodiversity (Van Nieuwenhuyse and LaPerriere 1986; McLeay et al. 1987; Gilvear et al. 1995). Placer and aggregate mining can be conducted on an industrial scale with heavy machinery (Fig. 3.7) or as artisanal mining with hand tools. Artisanal mining does not necessarily have a lesser impact on the river corridor, however, if it involves hundreds or even thousands of individuals working simultaneously in a concentrated portion of the river (Cordy et al. 2010).

Log floating refers to the use of channels to convey cut timber downstream to collection points or sawmills. Although this might sound relatively benign, log floating historically involved substantial modification of flow regime and channel boundaries. Log floating was conducted throughout river networks, including headwater channels barely wide enough to contain the diameter of a single log (Cowan 2003; McMahon and Karamanski 2009).

Splash dams modified flow regime. These temporary dams collected water and cut logs. Once filled, a splash dam was dynamited, sending an outburst flood and

Fig. 3.8 The Vindel River in northern Sweden was extensively historically modified for log float-ing. (**a**) 2008 aerial photograph in which linear structures remain from log floating. *Yellow arrow* indicates flow direction. (**b**) Ground photos before (*top*) and after (*lower row*) river restoration in 2004 along another portion of the Vindel. Upper row of ground-level photographs courtesy of Christer Nilsson. (Aerial view of the Vindel River near Mardsele courtesy of Google Earth)

mass of large wood downstream in a highly erosive pulse. More than a century after splash-damming ceased, splash-dammed channels in the western United States remain larger and simpler because of this erosion (Sedell et al. 1991; Young et al. 1994; Miller 2010). Channel boundaries were also directly modified to facilitate conveyance of the logs. Naturally occurring large wood, large boulders, and bedrock knobs were blasted out, and floodplains and secondary channels were blocked to prevent floating logs from becoming trapped in them (Fig. 3.8) (Young et al. 1994; Miller 2010). Log floating occurred wherever there was commercial timber harvest and commonly involved the movement of thousands of logs each year. Historical references describe these activities from the second century BC in the Italian Alps (Comiti 2012) and at least as far back as the fifteenth century in central, lowland Europe (Schama 1995). Log floating continued into the twentieth century in regions such as Siberia and northern Sweden (Nilsson et al. 2005b).

Floodplain drainage lowers the riparian water table, usually to facilitate cropping or other land uses in the floodplain. Drainage typically involves installing tile drains and cutting a grid of channels into the floodplain to convey water to the natural channel (Fig. 3.9). This activity, too, has a long history. The English invited Dutch engineers to help develop floodplain drainage during the seventeenth century, for example, and the library of early American agriculturalists such as George Washington included English manuals of land drainage (Simco et al. 2009). Floodplain drainage facilitates compaction and drying of floodplain soils, changing nutrient retention and habitat abundance. This type of drainage was typically undertaken in connection with other alterations of the river corridor, such as construction of levees and flow regulation (Wohl 2013).

3.2.2   Alterations of Fluxes of Materials and Movements of Organisms

This section briefly reviews levees and flow regulation as two particularly widespread, human-induced alterations of fluxes and connectivity within river corridors. Levees are naturally occurring depositional features at the boundary between the channel and floodplain. Levees form where flow velocity declines as peak flows move from the confined channel to the larger cross-sectional area and greater hydraulic resistance of the floodplain. Artificial levees are linear mounds built along the channel to completely or largely contain peak flows within the active channel. Artificial levees reduce or eliminate lateral connectivity within the river corridor, leading to drying and simplification of the floodplain. Artificial levees also result in greater velocity and erosive energy within the active channel during peak flows. This can enhance bed and bank erosion and downstream sedimentation. Levees also facilitate human settlement or development within flood-prone zones, exacerbating flood damages when the levees fail or are overtopped (Tobin 1995; Pinter 2005).

Flow regulation includes dams and diversions, although this section focuses on dams. Construction of dams has a long history: the first record of a dam comes from Jordan in 3000 BC (Smith 1971). The construction of large dams (> 15 m tall) and major dams (> 150 m tall) greatly accelerated during the latter half of the twentieth century (Gleick 2003) and numerous large dams continue to be built in Asia, Africa,

Fig. 3.9 Aerial view of the grid of tile drains installed to drain floodplains in the Le Sueur River basin of Minnesota, USA. (Ground and aerial photographs courtesy of Karen Gran, University of Minnesota, Duluth)

and Latin America. Large dams are no longer built in North America and Europe, partly because most of the suitable sites already have dams and partly because of the recognition of the enormous negative effects of dams on river ecosystems.

The primary effects of dams on river ecosystems are fourfold (Williams and Wolman 1984; Ligon et al. 1995; Kondolf 1997; Poff and Hart 2002) (Fig. 3.10). First, dams create large backwaters that store sediment and organic matter. This limits or eliminates supplies of these materials to downstream river segments, which can cause exacerbated downstream erosion and limit nutrient availability (Erskine 1985; Ligon et al. 1995; Wang et al. 2007a, b). Dams designed to be overtopped during peak flows, which are sometimes known as run-of-river dams, weirs, or barrages, can pass suspended sediment, but are still likely to store coarser sediment moving in contact with the river bed and thus create sediment deficits downstream.

Second, dams alter downstream flow regimes, typically by increasing base flows and decreasing peak flows (Magilligan and Nislow 2001), but also by creating abrupt diurnal fluctuations in flow in the case of hydropower dams. Reduction of peak flows causes multiple effects, including changes in channel geometry and planform such as narrowing, reduced lateral mobility, or transition from a braided to a single channel (Surian 1999; Swanson et al. 2011). Changes in channel form alter habitat abundance and distribution (Vietz et al. 2013) and reduced peak flows

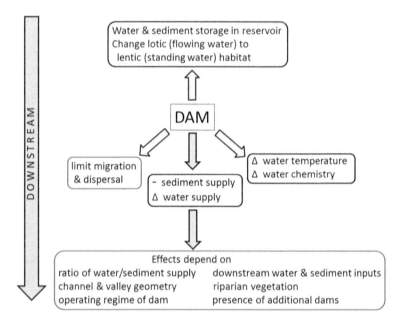

Fig. 3.10 Schematic illustration of the primary upstream and downstream effects of dams

also limit the ability of aquatic and riparian organisms to reproduce and disperse (Pringle et al. 2000). Smaller peak flows reduce the ability of the river to transport sediment that enters downstream from the dam, which can lead to a condition of excess sediment and channel aggradation (Salant et al. 2006; Ta et al. 2008). Finally, lower peak flows can limit lateral connectivity by eliminating processes such as overbank flooding (Walker and Thomas 1993; Renshaw et al. 2014). Abrupt fluctuations in flow associated with hydropower generation can strand and kill aquatic organisms such as juvenile fish along the channel margins (Stanford and Hauer 1992).

The key point to understand with respect to how dams alter water and sediment supplies to downstream portions of rivers is that changes in these primary input variables will result in changes to channel and floodplain form and process (Fig. 2.4). Changes to river form and process can in turn change the abundance and diversity of riverine organisms in undesirable ways (Nilsson and Berggren 2000; Bunn and Arthington 2002), as when habitat loss contributes to endangering a native fish species (Dudgeon 2000).

A third primary effect of dams is to block longitudinal migration and dispersal by aquatic organisms. Many dams have no provisions for aquatic organism passage and, where provisions such as fish ladders are present, these structures commonly do not fully restore longitudinal connectivity for the full diversity of aquatic organisms present in a river (Stanford and Hauer 1992; Liermann et al. 2012). Dams also commonly create a seed shadow downstream for plants that disperse their propagules via water transport (Jansson et al. 2000; Merritt and Wohl 2006; Burke et al. 2009).

The fourth primary effect of dams is to substantially alter water chemistry by changing water temperature, dissolved oxygen, and solute concentrations (Humborg et al. 1997; Bednarek and Hart 2005). Many dams only release water from the base of the dam. This water is likely to be colder and to have smaller suspended sediment concentrations than undammed flow on some rivers, which sometimes favors introduced aquatic species over native species (King et al. 1998; Clarkson and Childs 2000).

The details of how a particular dam affects upstream and downstream portions of a river depend on at least three factors (Ligon et al. 1995; Poff and Hart 2002). First of these is the physical and ecological characteristics of the river. Details such as tributary inputs downstream from the dam, the presence of erosionally resistant layers in the channel bed or banks (Jiongxin 1996), or the pre-dam form of the river (Musselman 2011) influence physical responses to the altered water and sediment regime below the dam. Ecological details such as the specific habitats required by downstream organisms, the migratory and reproductive patterns of those organisms, or the degree of endemism among organisms influence the ecological responses below the dam.

Second, the size and operating regime of the dam influence the dam's effects on the river ecosystem (Poff and Hart 2002). Dams can be operated primarily for water storage or for hydroelectric power generation, for example, which results in very different patterns of water release from the dam and differences in the degree to which flow regime downstream the dam is altered.

Finally, whether the dam is isolated or part of a series of dams along the river governs the dam's effects on the river ecosystem (Skalak et al. 2013). The effects of changes in water and sediment supply, in particular, vary with distance downstream from the dam. For the large, historically braided Missouri River in the USA, Skalak et al. (2013) document consistent downstream changes over distances of 100 to 140 km below each dam, but multiple dams with closer downstream spacing interrupt this spatial sequence.

A secondary effect of dams is that, by destabilizing the river corridor downstream, the dam can result in additional channel engineering. Sediment storage behind a dam can result in accelerated channel erosion, for example, leading to bed and bank stabilization as mitigation measures (Gendaszek et al. 2012).

Dam removal or changes to the operating regime of a dam can mitigate the negative effects created by the dam. Dam removal has been undertaken at many sites in the United States, with varying degrees of success in restoring river ecosystems (Bednarek 2001; Stanley and Doyle 2003). Although more than 1200 dams had been removed in the USA as of 2016, fewer than 10% of these removals have been scientifically evaluated and most evaluations involve only short-term (1–2 years) monitoring (Bellmore et al. 2016). Although restoration of longitudinal connectivity and a more natural flow regime may be immediate following dam removal, adjustments in sediment transport and channel and floodplain form both upstream and downstream from the former dam may continue for decades (Pizzuto 2002).

Changes to a dam's operating regime commonly involve one of three modifications. The first is releasing flows that mimic the natural hydrograph. This can involve experimental floods to mimic naturally occurring peak flows in order to restore particular channel forms or facilitate reproduction and dispersal of riverine organisms (Collier et al. 1997; Ortlepp and Mürle 2003; Wilcox and Shafroth 2013). Changes to a dam's operating regime can also focus on slowing the rate of change in discharge released by a hydropower dam so that aquatic organisms are able to move in response to changing flows (Meile et al. 2011). In some cases, a dam's operating regime is changed to create more fluctuations in base flow rather than maintaining a static discharge throughout the year (Auble et al. 1994).

A second common modification to dam operations is to increase sediment supply below the dam. Designing a dam to pass sediment is difficult and few dams are designed to pass sediment (Shen 1999). Instead, sediment can be bypassed around the dam or removed from the reservoir or elsewhere and dumped into the river downstream from the dam (Sumi and Kantoush 2010). Sediment bypassing or direct introduction below a dam tends to be expensive and it can be difficult to create a balance between altered water and sediment yields below a dam that results in desired physical and ecological effects.

A third basic type of modification to an existing dam involves modifying the dam to release water from different depths within the reservoir (Patten et al. 2001; Arthington and Pusey 2003; Shafroth et al. 2010). This improves the ability to regulate water temperature and chemistry below the dam.

3.2.3 Alterations of Riverine Organisms

This section reviews beaver trapping, introduced riverine species, commercial extinction of species, floodplain deforestation, and riparian grazing as five particularly common examples of human alterations of biotic communities in river ecosystems. Recognition of alterations of riverine biota is important in the context of river management for at least two reasons. First, removal of native species or introduction of non-native species can substantially alter the physical form of rivers, as when removal of beaver decreases floodplain wetlands or the introduction of non-native riparian trees changes bank stability and channel width. Second, the presence of non-native species may be the primary factor limiting recovery of native species that are the target of river management. An example discussed in more detail later in this chapter comes from the Colorado River in the United States, where the presence of introduced fish species severely limits the recovery of endemic fish species, even though habitat for endemic species is being restored.

Historical beaver trapping resulted from the fur trade, which caused near-extinction of *Castor fiber* in Eurasia and *Castor canadensis* in North America. Beaver continue to be killed, largely to prevent the backwaters associated with their dams or the mobilization of wood from their dams that can block downstream river infrastructure such as road culverts. The primary emphasis now throughout Europe

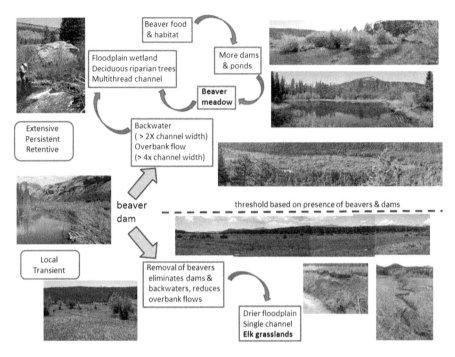

Fig. 3.11 Schematic illustration of the sequence of events that create a beaver meadow (upper portion of figure) in wide sections of river corridor when beaver are present and dams are maintained, versus the events that create an elk meadow (lower portion of figure) when beaver no longer maintain dams

(Nolet and Rosell 1998) and in many parts of the United States, however, is beaver restoration (Pollock et al. 2014; US FWS 2015).

Removal of beaver causes beaver dams to fall into disrepair. While present, beaver dams create backwaters and overbank flooding that slows downstream fluxes of water, sediment, nutrients, and organic matter (Naiman et al. 1994; Correll et al. 2000; Johnston 2014; Wegener et al. 2017). Beaver dams raise riparian water tables and increase complexity and lateral and vertical connectivity of the river corridor (John and Klein 2004; Polvi and Wohl 2012, 2014). Beaver dams and associated ponding of water also greatly enhance habitat diversity, biomass, and biodiversity (Pollock et al. 2003; Rosell et al. 2005; Wright 2009; Hood and Larson 2014) (Fig. 3.11).

When beaver are removed and dams fall into disrepair, peak flows are more likely to concentrate in a single channel rather than spreading among many smaller, subparallel channels. This results in greater flow velocity and erosive energy, leading to channel widening and incision. Erosion of the main channel commonly lowers the riparian water table. Loss of the beaver dams reduces complexity, connectivity, and retention within the river corridor (Green and Westbrook 2009).

Introduced riverine species are non-native plants and animals accidentally or deliberately introduced to a river corridor. Invasive introduced species are those that dramatically increase in population densities and geographic range after introduction. Introduced species can prey on or compete with native species for habitat and resources, leading to declines in native species (Townsend and Crowl 1991). Introduced species can also alter form and process within river ecosystems, as when introduced riparian vegetation stabilizes channel banks, traps sediment, and causes channel narrowing (Dean and Schmidt 2011), or introduced beavers in South America cause persistent flooding of riparian zones that kills native vegetation (Pastur et al. 2006).

Invasive species are of greatest concern, particularly if they cause physical and/ or biological changes in river ecosystems. Widely publicized examples of invasive introduced species in North America include the algae *Didymosphenia geminata*, which forms thick layers on stream beds and alters macroinvertebrate community structure and function (Gillis and Chalifour 2010), and zebra mussels (*Dreissena polymorpha*), which reach enormous numbers, consume large quantities of plankton and thus reduce food sources for other aquatic animals that consume plankton, and outcompete native filter-feeders such as bivalves (Strayer et al. 1999). Other invasive exotics in North America include bighead (*Hypophthalmichthys nobilis*) and silver carp (*H. molitrix*), which are Asian species that consume planktonic organisms on which most species of native fish in the Mississippi River basin depend (Chick and Pegg 2001). Among sport fisheries, introduced species include brown trout (*Salmo trutta*) native to Europe, brook trout (*Salvelinus fontinalis*) introduced to western North America from the eastern part of the continent, and rainbow trout (*Oncorhynchus mykiss*) introduced from the Pacific coast of North America to the Rocky Mountains and eastern North America. Each of these introduced trout species has the potential to alter consumption of aquatic and terrestrial insects and thus change food webs within river corridors (Baxter et al. 2007).

Invasive exotic species including willows (*Salix* spp.) (Cremer 2003), carp (*Cyprinus carpio*) (Shearer and Mulley 1978), and water hyacinth (*Eichhornia crassipes*) (Perna and Burrows 2005) have also strongly affected Australian river ecosystems. Although Eurasian and African river ecosystems also host non-native species, far more research and management have focused on non-native species in North America, Australia, and New Zealand (e.g., Townsend and Crowl 1991).

The invasive riparian woody plant tamarisk (*Tamarix* spp.) grows in such dense thickets along river banks and floodplains that it excludes many native riparian plants and alters flow resistance, bank stability, sedimentation, and channel form (Graf 1978; Patten 1998) (Fig. 3.12). Tamarisk exemplifies some of the uncertainties surrounding mitigation of invasive species. Among the concerns initially expressed regarding the spread of tamarisk in the United States were the potential for increased transpiration and associated declining stream flows and the potential loss of habitat for native songbirds and other riparian species. Massive campaigns using herbicide, cutting, and burning were undertaken to limit the spread of tamarisk (Hultine et al. 2010) and the tamarisk beetle (*Diorhabda elongata*) was introduced as a non-native biological control agent capable of tamarisk defoliation

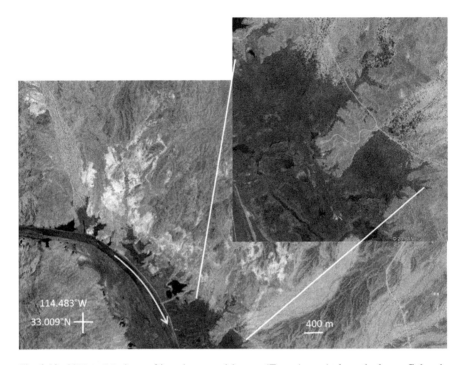

Fig. 3.12 2011 aerial views of invasive tamarisk trees (*Tamarix* spp.) along the lower Colorado River at Imperial National Wildlife Refuge near Yuma, Arizona, USA. The Colorado River at this location (flow direction indicated by *yellow arrow*) is channelized and highly regulated, with peak flows that reach only a small fraction of historical magnitudes. The ephemeral channels entering the Colorado at lower right in these views have extensive riparian tamarisk forests, which appear here as the *dark green* sections of the photographs. (Photos courtesy of Google Earth)

(Paxton et al. 2011). Subsequent research, however, indicated that tamarisk does not transpire nearly as much water as originally estimated and songbirds and other wildlife can successfully use tamarisk as habitat (Shafroth et al. 2005). This subsequent research calls into question the costs and environmental side effects of efforts to locally eradicate tamarisk.

Commercial extinction of species results from overharvest. Beaver are the prime example, as discussed above, but overharvest of invertebrates and fish can also significantly alter river corridors. North American examples include native unionid mussels and buffalo fish (*Ictiobus* spp.) in the Upper Mississippi River drainage.

Unionid mussels occur worldwide, but are particularly abundant and diverse in North America. These bottom-dwelling invertebrates process organic matter and mix streambed sediments in fine-grained channel substrates. Where large concentrations of mussels create mussel beds, their shells provide a firm substrate for other organisms such as insect eggs and larvae and fish eggs (Beckett et al. 1996; Morales et al. 2006). Mussels were commercially harvested from the late nineteenth century to the early twentieth century from the Upper Mississippi River basin for freshwater pearls and to cut disks from their shells for buttons (Blodgett et al. 1998). Harvest involved using a highly destructive instrument known as a crowfoot—a horizontal bar hung with hooks—that was dragged along the river bottom (Blodgett et al. 1998). Massive removal of mussels reduced abundance and diversity of aquatic habitat for other organisms and reduced organic matter processing in the stream bed. Because organic matter lowers the dissolved oxygen content of river water as it decays, removal of mussels indirectly altered water quality.

Buffalo fish are bottom-feeders that eat insect larvae, zooplankton, attached algae, shellfish, and organic detritus. These fish can grow to be more than a meter long and 37 kg in weight (Forbes 1988; Greenberg 2002). Buffalo fish supported the primary commercial fishery on the Illinois River, a tributary of the Mississippi, with up to ten million tons of the fish sent to commercial markets each year until they were fished to commercial extinction during the 1890s (Forbes and Richardson 1920; Greenberg 2002). Removal of these bottom-feeders from the river system likely altered water quality and aquatic food webs, although scientific studies documenting these effects were not conducted at the time of the population crash in these fish (Wohl 2013).

Floodplain deforestation is typically undertaken with floodplain drainage, levee construction, and other activities designed to facilitate human use of the floodplain. Removal of floodplain forests reduces the stability of the floodplain and the attenuation of downstream fluxes of water, solutes, sediment, and organic matter. Floodplain deforestation also reduces sources of large wood to the river corridor and thus habitat abundance and diversity in both the channel and floodplain. Reductions in habitat can lower biomass and biodiversity (Sedell and Froggatt 1984). Loss of shading and organic matter inputs also change nutrient availability and biogeochemical cycling.

Because the floodplain includes the riparian zone, floodplain deforestation results in loss of riparian buffers. Riparian buffers are vegetated bands along the active channel that filter fluxes of material from adjacent uplands and retain both nutrients and contaminants, thus improving water quality (Gregory et al. 1991; Naiman et al. 2005). Riparian buffers increase the resistance of the channel banks to erosion, thus limiting lateral channel movement and introduction of sediment from bank erosion (Griffin et al. 2005; Gurnell 2014). Riparian vegetation shades the channel, ameliorating diurnal and seasonal fluctuations in water temperature. Riparian vegetation also sheds organic matter into the channel, providing an important nutrient source for primary production (Naiman et al. 2005). Finally, riparian vegetation provides critical habitat for diverse organisms from invertebrates to mammals (Sabo et al. 2005). Floodplain deforestation and loss of riparian buffers thus create a cascade of secondary effects in river ecosystems.

Riparian grazing also reduces or eliminates riparian vegetation. Riparian grazing typically involves domestic animals, particularly cows. Wild ungulates that have reached unnaturally high population densities as a result of removal of predators can also severely overgraze riparian areas (Ripple and Beschta 2004).

Sustained, high-intensity riparian grazing can create a variety of effects (Trimble and Mendel 1995; Myers and Swanson 1996). These include reducing or eliminating favored plant species. Continued high numbers of grazing animals compact floodplain soils, which reduces infiltration and increases runoff and erosion. Following established trails, grazing animals break down banks and enhance sediment yields to channels. As woody riparian vegetation density declines, channel margins are less shaded, which influences water temperature and photosynthesis within the channel. Grazing animals add excess nutrients to rivers through their wastes. Finally, bank trampling and removal of woody vegetation result in the formation of relatively wide, shallow channels with mobile beds of fine-grained sediment and a lack of pools, all of which limits habitat abundance and stability for stream organisms. Numerous studies tie riparian grazing to reduced water quality and declines in macroinvertebrate and fish populations (e.g., Kauffman and Krueger 1984; Belsky et al. 1999).

3.3   Cumulative Effects

Although the previous section reviews the effects on river ecosystems of individual forms of land and resource use, an important consideration is that each type of human alteration rarely occurs in isolation. In most altered river basins, multiple human activities have occurred simultaneously or in succession through time. These activities tend to be particularly well documented in regions that experienced intensive resource use within the past two to three centuries, such as North America, Australia, and New Zealand, but the combined effects of human-induced changes now characterize most rivers around the world (Nilsson et al. 2005a).

The predominant cumulative effect of indirect and direct alteration of river ecosystems is to simplify and homogenize water and sediment inputs, physical form, and biota (Moyle and Mount 2007; Peipoch et al. 2015). As noted in the first chapter, the secondary effects of this simplification and homogenization show up in loss of freshwater species, declines in water quality and quantity, flood hazards, exacerbated channel erosion, eutrophication, and coastal erosion. Four major drainage basins can be used to illustrate these cumulative effects: the Mississippi River of North America; the Murray-Darling River of Australia; the Danube River of Europe; and the Huanghe (Yellow River) of Asia.

3.3.1   The Mississippi River Drainage Basin of North America

The Mississippi River drains 3.5 million square kilometers of North America, predominantly within the United States (Fig. 3.13). The majority of the river's water flow comes from the central and eastern half of the drainage via the Ohio, Illinois, and Upper Mississippi Rivers. The majority of the sediment discharges comes from the Missouri River in the western half of the drainage (Meade and Moody 2010). The larger rivers within the drainage basin have been used for navigation for centuries. Modification of the rivers to enhance navigation started with commercial

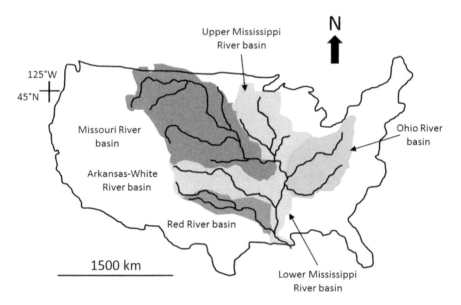

Fig. 3.13 The Mississippi River basin within the conterminous United States. Different sub-basins and primary rivers indicated

steamboat traffic, primarily during the early decades of the nineteenth century (Wohl 2014), and continues at present as part of the extensive transportation network for barges (Wohl 2011). Millions of naturally occurring pieces of large wood were removed from the river and many millions more of cut logs were rafted downstream to collection booms and sawmills (Wohl 2014). The river was dredged, straightened, channelized, and leveed.

Longitudinal connectivity within the drainage basin is highly reduced as a result of eight major dams on the Missouri River, 20 locks and dams on the Upper Mississippi (Fig. 3.14), 21 locks and dams on the tributary Ohio River, eight locks and dams on the tributary Illinois River, nine locks and dams on the tributary Tennessee River, and so on through a long list of substantial tributaries. Water and sediment movements in the river network of the Mississippi River drainage are highly regulated and bear little resemblance to natural fluxes (Knox 2007). Before 1900, the Mississippi transported an estimated 400 million metric tons of suspended sediment to the coast, but by 1987–2006 transport averaged 145 million metric tons per year (Meade and Moody 2010). By 2011, median sediment load just above the river delta was 73% below the load carried before all of the dams were built (Heimann et al. 2011). About half of the decline results from sediment trapping behind dams, with the remainder resulting from the combined effects of channel stabilization and upland soil erosion controls (Meade and Moody 2010).

The Mississippi drainage also hosts some of the most extensive agricultural lands in North America and receives associated pesticides and fertilizers in runoff from these lands, as well as several major urban and industrial areas that contribute

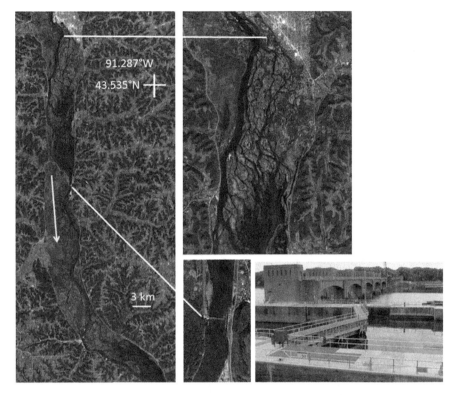

Fig. 3.14 Aerial views of a lock and dam and upstream river corridor on the Upper Mississippi River near Genoa, Wisconsin, USA in 2015. Photo at upper right indicates more natural segment of river upstream from the lock and dam, with multiple secondary channels and backwaters. Aerial photo at lower right is closer view of lock and dam; ground photo at lower right is another lock and dam along the Upper Mississippi. (Aerial photographs courtesy of Google Earth)

a diverse array of pollutants to the river network (Wohl 2004). Combined with widespread drainage of floodplain wetlands and removal of floodplain forests and riparian buffer strips, the river's capacity to metabolize nutrients and store organic carbon and other dissolved and particulate materials has been severely reduced. Hanberry et al. (2015) estimate that the lower Mississippi River alluvial valley historically stored 234 Tg of organic carbon in floodplain forest vegetation and soils, whereas contemporary storage is estimated at 97 Tg.

In addition to reduced storage of organic carbon, the cumulative effects of this long history of intensive and spatially extensive alteration of the Mississippi River drainage basin manifests in several ways. In the World Wildlife Fund's 2013 assessment of the conservation status of freshwater ecoregions in North America, the central upper and lower portions of the basin are rated as endangered overall, with greater than 50% of the catchment affected by altered land cover, degraded surface water quality, altered flow regime, and very high habitat fragmentation (WWF 2013). This is especially unfortunate given that much of the ecoregion ranks highly with respect to total species richness and total number of endemic species (WWF 2013).

With regard to water quality, the Mississippi River drainage is in equally bad shape. Levels of more than 80 pesticides and pesticide metabolites tested for by the U.S. Geological Survey routinely exceed the guidelines for sustaining aquatic life, particularly in rivers draining urban areas. Other pollutants include additional organochlorine compounds (e.g., PCBs), trace elements (e.g., arsenic, mercury, cadmium), volatile organic compounds (e.g., toluene, MTBE), and pathogens (Kleiss et al. 2000). These are among the factors underlying the informal designation of the lower Mississippi River region between Baton Rouge and New Orleans, Louisiana as 'cancer alley.'

Like an unfortunately large number of other rivers around the world, the Mississippi is now infamous for the presence of an extensive hypoxic or dead zone around its mouth. Hypoxia occurs when dissolved oxygen levels are ≤ 2 mg/L (Rabalais et al. 2007). This results from excess nutrients that lead to high primary production and then decay of organic matter. The Mississippi and the nearby Atchafalaya River deliver 91% of the annual nitrogen load and 88% of the annual phosphorus load to the northern Gulf of Mexico (Dunn 1996). The northern Gulf water is stratified as a result of salinity and thermal differences and this stratification intensifies during summer months with thermal warming of surface waters (Wiseman et al. 1997). Nutrient delivery from the rivers fuels high organic production in the surface waters, creating a flux of carbon to the bottom waters in the form of senescent plankton and fecal pellets. Decomposition of this material by aerobic bacteria consumes dissolved oxygen faster than it is resupplied from stratified surface waters, causing hypoxia that persists for months in bottom waters (CENR 2000; Rabalais et al. 2007). Few marine animals can survive in hypoxic conditions, which can extend upward into the water column (Rabalais and Turner 2001). The first systematic mapping and monitoring of dissolved oxygen levels around the mouth of the Mississippi began in 1985 and revealed the existence of a hypoxic zone in the Gulf of Mexico. Between 1985 and 1992 the extent of this zone averaged 8200 km^2, but the average extent increased to 15,900 km^2 between 1992 and 2007, reaching a historic maximum extent of 22,000 km^2 in 2002 (Rabalais et al. 2007).

Declining sediment delivery to the mouth of the Mississippi River has caused delta and coastal erosion. The Mississippi reaches the Gulf of Mexico at the border of the State of Louisiana. This state has been losing coastal wetlands at a rate that starts at 17 km^2 each year in 1917 and rises to 117 km^2 a year during the 1960s before leveling off to the present rate of 69 km^2 pear year (Boesch et al. 1994). In addition to declining sediment delivery by the river, oil and gas pipelines and access canals within the coastal wetlands create pollution and salinity intrusions that kill wetland vegetation. When present, this vegetation stabilizes barrier islands off the river's mouth (Turner 1997). Vegetation die offs exacerbate coastal erosion. Vegetation in inland coastal wetlands is also stressed by lack of flooding and associated nutrient delivery as a result of levees and flow regulation, as well as herbivory by nutria (*Myocastor coypus*), a large rodent introduced from South America. Sea level at the Mississippi delta is rising by 1.2–4.3 cm per year. This rise results from the combined effects of glacial melting elsewhere in the world; subsidence of coastal sediments because of removal of water, oil, and natural gas; and lack of continued deposition of new sediment (Blum and Roberts 2009).

In summary, the cumulative effects of upland alteration, river engineering, flow regulation, contaminants, and introduced species within the Mississippi River drainage basin result in high rates of extinction for native freshwater species; the presence of an enormous dead zone in the coastal waters receiving the river's discharge; contaminated water; a shrinking delta; and a plethora of inter-related changes to tributary rivers throughout the drainage basin. Attempts to reverse or mitigate these trends through management and restoration programs focused on distinct portions of the river basin and delta together run into the billions of dollars.

3.3.2 Australia's Murray-Darling River Drainage Basin

The Murray and Darling Rivers together drain nearly 1.1 million square kilometers of southeastern Australia (Fig. 3.15). The Darling drains a larger area but the Murray contributes more discharge at the point where the rivers join before flowing into the Pacific Ocean. Neither river, however, has much flow by world standards. The entire basin produces an average flow of less than 800 cubic meters per second, compared to nearly 17,000 for the Mississippi or 7000 for the Danube. The history of human alteration of the Murray-Darling drainage basin exemplifies that of many of dryland rivers around the world.

Like the Mississippi, the Murray-Darling underwent intensive human alteration relatively recently compared to many rivers in Eurasia. Europeans explored the drainage basin during the first half of the nineteenth century and intensive use of natural resources quickly followed. Much of the Murray-Darling drainage is grass-land or desert, but extensive riparian forests dominated by river red gum (*Eucalyptus camaldulensis*) were historically present. These forests were cut to expand grazing and crop lands and to provide fuel for steamboats (Fig. 3.16). Large wood was removed from the river to enhance navigation and reduce flooding. Water was diverted for irrigated agriculture starting during the 1860s. Diversions accelerated through time, along with construction of locks, dams, and large reservoirs, and channelization of the river (Wohl 2011).

Among the primary concerns in the Murray-Darling today are species loss, intro-duced species, salinization, and algal blooms (Walker and Thoms 1993). The drain-age basin historically contained 34 fish species, half of which are now threatened as a result of habitat loss, declining water quality and quantity, overfishing, and intro-duced species such as carp (Fisher 1996). Among the primary introduced species that affect river process and form is willow (*Salix* spp.). Willows were planted along the main channel of the Murray to distinguish it from secondary channels as part of navigation improvements and to stabilize the banks of irrigation canals across the floodplain. The willows have become invasive, excluding native vegetation. The

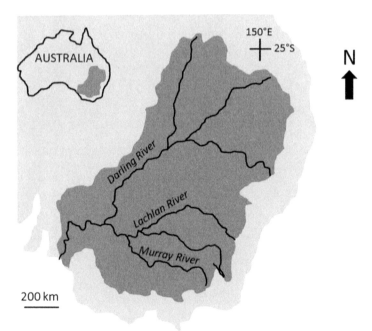

Fig. 3.15 The Murray-Darling River basin (shaded in *brown*) of southeastern Australia, with major rivers labeled. Inset map of Australia shows location of river basin within the continent

willows provide poor habitat for native animals and reduce the recruitment to channels of large wood that provides vital habitat for native fish. Willows also alter aquatic and riparian food webs through shading of river margins and changes in leaf litter volume and nutrient content of leaf detritus (Schulze and Walker 1997).

Salinization of river waters and floodplain soils in the Murray-Darling drainage results from three factors. The first is naturally high evaporation rates in the dry climate. The second and third factors reflect human alterations. Mobilization of salts from soils as irrigation water infiltrates creates salinization, as does replacement of native vegetation with introduced plants. The introduced plants have lower transpiration rates. This allows riparian water tables to rise and naturally saline ground water to reach the surface (Allison et al. 1990). Salinization is of concern because it limits growth of native vegetation and crops in river bottomlands and can exacerbate saline runoff into river channels.

Increased nutrient runoff to channels has exacerbated blooms of blue-green algae in river waters of the Murray-Darling drainage. Low volumes of slowly moving flow associated with water withdrawals from the river network and ponds upstream of locks and dams also exacerbate algal blooms (Bowling and Baker 1996). Blue-green algal blooms create problems because they can produce toxic byproducts that sicken or kill other organisms, including humans. An algal bloom extending along 1000 km of the Darling River in 1991 killed an estimated 1600 sheep and cattle and threatened human water supplies for several rural regions (Wohl 2011).

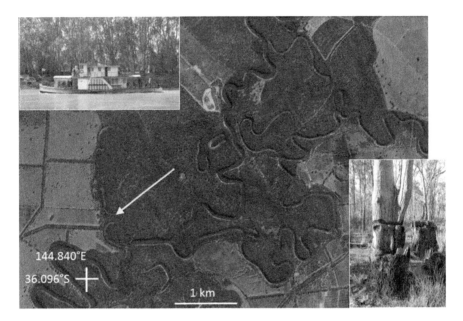

Fig. 3.16 The Barmah floodplain forest along the Murray River is one of the most extensive remaining, as seen in this 2002 aerial view of the river corridor. Inset photos show a steamboat, now operated for tourists, and river red gum trees in the Barmah that were ring-barked decades before this 2016 photo was taken, in order to kill the trees and increase grazing land on the floodplain. (Aerial photograph courtesy of Google Earth)

As in the Mississippi River drainage basin, growing alarm over these cumulative effects in the Murray-Darling drainage has led to multiple management and restoration projects focused on specific portions of the river network and the drainage basin as a whole (Arthington and Pusey 2003; Brooks and Lake 2007). In some respects, the sense of urgency underlying these projects may be greater in Australia than in the United States. Australia has experienced rapid population growth through immigration in recent decades and potable water supplies are severely limited by a dry climate that is likely to experience more intense, frequent, and prolonged droughts under a warming climate (Hughes 2003).

3.3.3 The Danube River Drainage Basin of Europe

The Danube River (Fig. 3.17) forms the premier east-west navigational corridor in central Europe because the river flows from headwaters in Germany and eventually enters the Black Sea via Romania (Wohl 2011). En route, the river crosses multiple highlands and intervening alluvial basins and drains an area of 817,000 km^2. As might be expected, the river network has a long history of channelization (Hohensinner et al. 2004), levee construction, flood control, construction of dams

Fig. 3.17 The Danube River basin (shaded in *gray*) and the borders of the primary countries within the basin. Inset map shows the location of the basin within a larger portion of Europe

for hydroelectric power and for navigation (Fitzmaurice 1996), construction of navigational canals (e.g., the Rhine-Main-Danube Canal completed in 1992; Lóczy 2007), and water pollution. Because the drainage basin includes portions of the territories of 18 nations, coordinating river management has been a challenging international undertaking (Nachtnebel 2000; Bloesch and Sieber 2003). Among the primary issues in the Danube drainage today are channelization, flow regulation, and pollution.

Channelization begins in alpine headwater tributaries that are 'trained' to confine braided planforms to a single channel and to limit bank erosion and bed incision (Fig. 3.18a). Channelization continues throughout the Danube drainage basin to the delta, where one of the three main distributary channels has been dredged and channelized for shipping since 1857 (Wohl 2011). Part of the river training in mountainous portions of the drainage basin includes check dams and sediment detention basins designed to limit downstream transmission of sediment during debris flows and floods (Fig. 3.18b). This sediment retention, along with stabilized banks in downstream portions of the river network, creates sediment deficits that exacerbate bed and bank erosion, leading to additional stabilization of the channel boundaries. Most of the portions of the Danube and its tributaries flowing across alluvial basins are now highly channelized (Fig. 3.18c) and flow in these river segments and in the confined canyons between alluvial basins is fully regulated by dozens of dams.

Most of the dams within the Danube drainage basin are designed for hydropower, navigation, and flood control, rather than for water storage. Consequently, many of the dams are relatively small, although they are extremely numerous (Fig. 3.19). A key concern with these dams is that they limit flood magnitude and

Fig. 3.18 (a) 2015 aerial view of the Austrian alpine town of Sankt Jakob in Defereggen. Note the uniform width of the channel at the bottom of the photo and the regularly spaced white patches along the tributary entering from the upper part of the photo; each of these white patches is aerated flow passing over a grade-control structure. *Yellow arrows* indicate flow direction in each channel. Ground photo of the channelized stream. (Aerial photo courtesy of Google Earth) (**b**) A check dam filled with sediment along the Hopfgartnergraben, a small, steep mountain channel near Sankt Jakob. Photo at left shows a closer view of the check dam, with people for scale; photo at right shows the larger context. (**c**) Aerial views of the Danube at Linz, Austria in 2015, illustrating a highly channelized portion of the river with minimal natural floodplain (*left*) and in the Alluvial National Park downstream from Vienna in 2016 (*right*). Although the Danube is unnaturally uniform and straight in the national park, the active channel is bordered by a forested floodplain that includes secondary channels. (Photographs courtesy of Google Earth)

c

Fig 3.18 (continued)

Fig. 3.19 Hydropower dam on the Danube River near Jochenstein, Germany. The river here drains approximately 50,000 km² and is approximately 400 m wide at the dam

associated lateral connectivity within the river corridor, thus limiting the effectiveness of efforts to reconnect the main channel with secondary channels and the floodplain along portions of the Danube. There are also some very large dams within the Danube, particularly the Djerdap I and II dams in Romania, built in 1970 and 1984, respectively, for hydropower (Wohl 2011). These dams effectively ended the migration of fish between the upper and lower portions of the Danube river network.

Among the affected fish are commercially important species such as beluga (*Huso huso*), Russian sturgeon (*Acipenser gueldenstaedtii*), and stellate sturgeon (*A. stellatus*) (Lenhardt et al. 2004).

Better sewage treatment and strict water quality standards have improved water quality in the upper and middle portions of the Danube markedly since the 1970s. Pollution remains severe, however, in the lower Danube drainage basin because of limited sewage treatment, agricultural runoff, and industrial discharges (Lucas 2001; Damacija et al. 2003).

River management across the Danube River drainage now focuses on three activities. The first is reconnecting main channels with secondary channels and floodplain areas through removal of physical barriers and alteration of flow regimes below dams (Tockner et al. 1998; Chovanec et al. 2002; Aarts et al. 2004). The second is restoring some channel complexity and sediment dynamics along highly channelized portions of the tributaries and upper river (Habersack and Piégay 2008; Muhar et al. 2008). Finally, management emphasizes restoring a more natural flow regime and improving water quality in the lower basin. Although the international character of the Danube drainage basin creates some unique challenges relative to the Mississippi or Murray-Darling examples, coordinated river basin management is now a priority within Europe.

3.3.4 The Huanghe Drainage Basin of China

The Huanghe (Yellow River; Fig. 3.20) has been nicknamed 'China's Sorrow' because of the long history of damaging floods associated with the river network. Among the salient characteristics of the Huanghe are its length (at 5460 km long, it is the second longest river in Asia) and its history of enormous floods. A flood in the fourteenth century, for example, killed an estimated seven million people. The Huanghe is also one of a growing number of rivers around the world that has recently (since 1972) shown a tendency to completely dry out before reaching the ocean because of consumptive water use (Yu 2002).

The Huanghe is also known for high concentrations of suspended sediment dominated by silt eroded from the Loess Plateau. High suspended sediment levels have contributed to sediment deposition that raised the bed of the channel through time, to which people responded by continuing to increase the height of the artificial levees along the channel. As a result, the elevation of the channel is now higher than the surrounding floodplain along parts of the river's course.

The Huanghe had high suspended sediment concentrations prior to human alteration of the watershed, but human activities have caused fine sediment inputs to and transport within the river network to fluctuate substantially through time. These fluctuations have resulted from historic clearance of native upland vegetation; construction of large reservoirs starting in the 1950s; soil conservation practices since

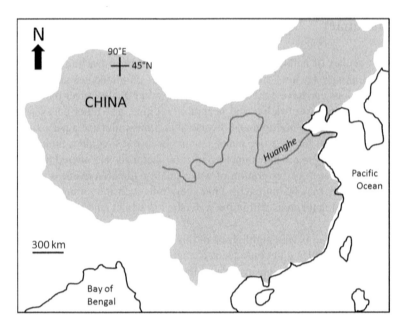

Fig. 3.20 Map of the Huanghe drainage basin (shaded in *orange*), showing the main stem of the river and the extent of China (shaded in *gray*)

the late 1970s; and, since the 1970s, water consumption that reduces river flow (Wang et al. 2007a, b). By the start of the twenty-first century, Miao et al. (2011) estimate that human activities were responsible for 55% of the reductions in water discharge and 54% of the reduced sediment load in the basin observed since 1970, with the remaining portions attributed to climate change.

Reduced water and sediment discharges to the Huanghe delta, along with oil drilling and associated development, have caused substantial salinization, wetland degradation, and delta erosion. Restoration of delta wetlands within nature reserves on the delta began in 2002, when altered water management within the river basin stabilized river discharge to the delta (Cui et al. 2009). Water quality, soil fertility, habitat, and bird diversity have all increased as a result of restoration activities, although delta erosion remains a serious problem (Peng et al. 2010). As with other river drainage basins, the cumulative effects across the watershed and throughout the channel network have resulted in severe alterations of the Huanghe River delta, which historically was a dynamic depositional area that buffered coastal regions from storm surges and supported high levels of biodiversity and agricultural productivity (Wang et al. 2012) (Figs. 3.19 and 3.20).

3.4   Summary

The brief reviews of conditions in the Mississippi, Murray-Darling, Danube, and Huanghe drainage basins illustrate how multiple simultaneous and prolonged human alterations can have cumulative effects on water quality, river biota, the stability and productivity of deltas and nearshore regions, and other ecosystem services. Recognizing that multiple and diverse alterations influence a particular river ecosystem is critical for effective management because it facilitates understanding that, although management may target a particular factor such as altered flow regime for a desired end such as restoration of water quality, the effectiveness of this targeted management may be limited by other alterations such as continued high inputs of nitrates from upland areas within the watershed in which land cover was historically altered.

The four drainage basins highlighted in this chapter are not unique in their history of human activities and consequent changes in river ecosystems. As of 2005, only a few of the large river basins around the world were relatively unaffected by flow regulation. These rivers, which tend to be in high- and low-latitude regions rather than in the temperate latitudes (Nilsson et al. 2005a), are in many cases slated for extensive development. At the same time, however, efforts to restore and maintain rivers as ecosystems are increasing within the temperate-latitude river basins that have already been highly altered by people. The next chapter reviews some of these efforts to restore rivers as highly functional ecosystems.

References

Aarts BGW, van den Brink FWB, Nienhuis PH (2004) Habitat loss as the main cause of the slow recovery of fish faunas of regulated large rivers in Europe: the transversal floodplain gradient. River Res Appl 20:3–23

Allison GB, Cook PG, Barnett SR, Walker GR, Jolly ID, Hughes MW (1990) Land clearance and river salinization in the western Murray basin, Australia. J Hydrol 119:1–20

Arthington AH, Pusey BJ (2003) Flow restoration and protection in Australian rivers. River Res Appl 19:377–395

Auble GT, Friedman JM, Scott ML (1994) Relating riparian vegetation to present and future streamflows. Ecol Appl 4:544–554

Baillie BR, Davies TR (2002) Influence of large woody debris on channel morphology in native forest and pine plantation streams in the Nelson region, New Zealand. N Z J Mar Freshw Res 36:763–774

Baker DW, Bledsoe BP, Albano CM, Poff NL (2011) Downstream effects of diversion dams on sediment and hydraulic conditions of Rocky Mountain streams. River Res Appl 27:388–401

Baxter CV, Fausch KD, Murakami M, Chapman PL (2007) Invading rainbow trout usurp a terrestrial prey subsidy from native charr and reduce their growth and abundance. Oecologia 153:461–470

Beckett DC, Green BW, Thomas SA, Miller AC (1996) Epizoic invertebrate communities on upper Mississippi River unionid bivalves. Am Midl Nat 135:102–114

Bednarek AT (2001) Undamming rivers: a review of the ecological impacts of dam removal. Environ Manag 27:803–814

Bednarek AT, Hart DD (2005) Modifying dam operations to restore rivers: ecological responses to Tennessee River dam mitigation. Ecol Appl 15:997–1008

Beechie T, Imaki H, Greene J, Wade A, Wu H, Pess G, Roni P, Kimball J, Stanford J, Kiffney P, Mantua N (2013) Restoring salmon habitat for a changing climate. River Res Appl 29:939–960

Bellmore JR, Duda JJ, Craig LS, Greene SL, Torgersen CE, Collins MJ, Vittum K (2016) Status and trends of dam removal research in the United States. WIREs Water. doi:10.1002/war2.1164

Belsky AJ, Matzke A, Uselman S (1999) Survey of livestock influences on stream and riparian ecosystems in the western United States. J Soil Water Conserv 54:419–431

Blakely TJ, Harding JS, McIntosh AR, Winterbourn MJ (2006) Barriers to the recovery of aquatic insect communities in urban streams. Freshw Biol 51:1634–1645

Bledsoe BP, Watson CC (2001) Effects of urbanization on channel instability. J Am Water Resour Assoc 37:255–270

Blodgett KD, Sparks RE, Whitney SD, Williamson R (1998) Mussel resources of the Illinois River system: values to Illinois' economy and natural heritage. In: Long term resource monitoring program 98-R012. US Geological Survey, Environmental Management Technical Center, Onalaska, WI

Bloesch J, Sieber U (2003) The morphological destruction and subsequent restoration programmes of large rivers in Europe. Arch Hydrobiol Suppl 147:363–385

Blum MD, Roberts HH (2009) Drowning of the Mississippi Delta due to insufficient sediment supply and global sea-level rise. Nat Geosci 2:488–491

Boesch DF, Josselyn MN, Mehta AJ, Morris JT, Nuttle WK, Simenstad CA, Swift DJP (1994) Scientific assessment of coastal wetland loss, restoration and management in Louisiana. J Coast Res Special Issue 20:i–v

Bowling LC, Baker PD (1996) Major cyanobacterial bloom in the Barwon-Darling River, Australia, in 1991, and underlying limnological conditions. Mar Freshw Res 47:643–657

Brooks AP, Brierley GJ (1997) Geomorphic responses of lower Bega River to catchment disturbance, 1851-1926. Geomorphology 18:291–304

Brooks SS, Lake PS (2007) River restoration in Victoria, Australia: change is in the wind, and none too soon. Restor Ecol 15:584–591

Bunn SE, Arthington AH (2002) Basic principles and ecological consequences of altered flow regimes for aquatic biodiversity. Environ Manag 30:492–507

Burke M, Jorde K, Buffington JM (2009) Application of a hierarchical framework for assessing environmental impacts of dam operation: changes in streamflow, bed mobility and recruitment of riparian trees in a western north American river. J Environ Manag 90:5224–5236

CENR (Committee on Environment and Natural Resources) (2000) Integrated assessment of hypoxia in the northern Gulf of Mexico. National Science and Technology Council, Washington, DC

Chick JH, Pegg MA (2001) Invasive carp in the Mississippi River basin. Science 292:2250–2251

Chovanec A, Schiemer F, Waidbacher H, Spolwind R (2002) Rehabilitation of a heavily modified river section of the Danube in Vienna (Austria): biological assessment of landscape linkages on different scales. Int Rev Hydrobiol 87:183–195

Clarkson RW, Childs MR (2000) Temperature effects of hypolimnial-release dams on early life stage of Colorado River basin big-river fishes. Copeia 2000:402–412

Collier MP, Webb RH, Andrews ED (1997) Experimental flooding grand canyon. Sci Am 276:66–73

Collins BD, Montgomery DR, Haas AD (2002) Historical changes in the distribution and functions of large wood in Puget lowland rivers. Can J Fish Aquat Sci 59:66–76

Comiti F (2012) How natural are alpine mountain rivers? Evidence from the Italian alps. Earth Surf Process Landf 37:693–707

Cordy P, Veiga MM, Salih I, Al-Saadi S, Console S, Garcia O, Mesa LA, Velasquez-Lopez PC, Roeser M (2010) Mercury contamination from artisanal gold mining in Antioquia, Colombia: the world's highest per capita mercury pollution. Sci Total Environ 410–411:154–160

Correll DL, Jordan TE, Weller DE (2000) Beaver pond biogeochemical effects in the Maryland Coastal Plains. Biogeochemistry 49:217–239

Cowan MM (2003) Timberrr!: a history of logging in New England. The Millbrook Press, Brookfield, CT

Cremer KW (2003) Introduced willows can become invasive pests in Australia. Biodiversity 4:17–24

Cui B, Yang Q, Yang Z, Zhang K (2009) Evaluating the ecological performance of wetland restoration in the Yellow River Delta, China. Ecol Eng 35:1090–1103

Czech W, Radecki-Pawlik A, Wyzga B, Hajdukiewicz H (2015) Modelling the flood capacity of a polish Carpathian river: a comparison of constrained and free channel conditions. Geomorphology 272:32–42

Damacija B, Ivancev-Tumbas I, Zejak J, Djurendic M (2003) Case study of petroleum contaminated area of Novi Sad after NATO bombing in Yugoslavia. Soil Sediment Contam 12:591–611

Davies PM (2010) Climate change implications for river restoration in global biodiversity hotspots. Restor Ecol 18:261–268

De Boer DH (1997) Changing contributions of suspended sediment sources in small basins resulting from European settlement on the Canadian prairies. Earth Surf Process Landf 22:623–639

Dean DJ, Schmidt JC (2011) The role of feedback mechanisms in historic channel changes of the lower Rio Grande in the big bend region. Geomorphology 126:333–349

DeWalle DR, Swistock BR, Johnson TE, McGuire KJ (2000) Potential effects of climate change and urbanization on mean annual streamflow in the United States. Water Resour Res 36:2655–2664

Dudgeon D (2000) Large-scale hydrological changes in tropical Asia: prospects for riverine biodiversity. Bioscience 50:793–806

Dudko S, Adriano DC (1995) Environmental impacts of metal ore processing and mining: a review. J Environ Qual 26:590–602

Dunn DD (1996) Trends in nutrient inflows to the Gulf of Mexico from streams draining the conterminous United States 1972–1993. US Geological Survey Water-Resources Investigations Report 96–4113

DuPont (1925) Ditching with dynamite. E.I. DuPont De Nemours, Wilmington, DE

Erskine WD (1985) Downstream geomorphic impacts of large dams: the case of Glenbawn dam, NSW. Appl Geogr 5:195–210

Evans BF, Townsend CR, Crowl TA (1993) Distribution and abundance of coarse woody debris in some southern New Zealand streams from contrasting forest catchments. N Z J Mar Freshw Res 27:227–239

Fisher T (1996) Fish out of water: the plight of native fish in the Murray-darling. Fish Fisheries Worldwide 24:17–24

Fitzmaurice J (1996) Damming the Danube: Gabčikovo and post-communist politics in Europe. Westview Press, Boulder, CO

Forbes SA (1988) On the food relations of fresh-water fishes: a summary and discussion. Illinois State Lab Nat Hist Bull 2(7):433–473

Forbes SA, Richardson RE (1920) The fishes of Illinois. State of Illinois, Springfield, IL

Ford CR, Vose JM (2007) *Tsuga canadensis* (L.) carr. Mortality will impact hydrologic processes in southern Appalachian forest ecosystems. Ecol Appl 17:1156–1167

Fransen PJB, Phillips CJ, Fahey BD (2001) Forest road erosion in New Zealand: overview. Earth Surf Process Landf 26:165–174

Frings RM, Berbee BM, Erkens G, Kleinhans MG, Gouw MJP (2009) Human-induced changes in bed shear stress and bed grain size in the river Waal (The Netherlands) during the past 900 years. Earth Surf Process Landf 34:503–514

Gabbud C, Lane SN (2015) Ecosystem impacts of alpine water intakes for hydropower: the challenge of sediment management. WIREs Water. doi:10.1002/wat2.1124

Gendaszek AS, Magirl CS, Czuba CR (2012) Geomorphic response to flow regulation and channel and floodplain alteration in the gravel-bedded Cedar River, Washington, USA. Geomorphology 179:258–268

Gillette R (1972) Stream channelization: conflict between ditchers, conservationists. Science 176:890–894

Gillis C-A, Chalifour M (2010) Changes in the macrobenthic community structure following the introduction of the invasive algae *Didymosphenia geminata* in the Matapedia River (Québec, Canada). Hydrobiologia 647:63–70

Gilvear DJ, Waters TM, Milner AM (1995) Image analysis of aerial photography to quantify changes in channel morphology and instream habitat following placer mining in interior Alaska. Freshw Biol 34:389–398

Gleick PH (2003) Global freshwater resources: soft-path solutions for the 21st century. Science 302:1524–1528

Goode JR, Luce CH, Buffington JM (2012) Enhanced sediment delivery in a changing climate in semi-arid mountain basins: implications for water resource management and aquatic habitat in the northern Rocky Mountains. Geomorphology 139-140:1–15

Graf WL (1978) Fluvial adjustments to the spread of tamarisk in the Colorado plateau region. Geol Soc Am Bull 89:1491–1501

Green KC, Westbrook CJ (2009) Changes in riparian area structure, channel hydraulics, and sediment yield following loss of beaver dams, BC. J Ecosyst Manage 10:68–79

Greenberg J (2002) A natural history of the Chicago region. University of Chicago Press, Chicago, IL

Greenwood H, O'Dowd DJ, Lake PS (2004) Willow (*Salix x rubens*) invasion of the riparian zone in south-eastern Australia: reduced abundance and altered composition of terrestrial arthropods. Divers Distrib 10:485–492

Gregory SV, Swanson FJ, McKee WA, Cummins KW (1991) An ecosystem perspective of riparian zones. Bioscience 41:540–551

Griffin ER, Kean JW, Vincent KR, Smith JD, Friedman JM (2005) Modeling effects of bank friction and woody bank vegetation on channel flow and boundary shear stress in the Rio Puerco, New Mexico. J Geophys Res 110:F04023. doi:10.1029/2005JF000322

Gurnell A (2014) Plants as river system engineers. Earth Surf Process Landf 39:4–25

Habersack H, Piégay H (2008) River restoration in the alps and their surroundings: past experience and future challenges. In: Habersack H, Piégay H, Rinaldi M (eds) Gravel-bed rivers VI: from process understanding to river restoration. Elsevier, Amsterdam, pp 703–737

Haidvogl G, Pont D, Dolak H, Hohensinner S (2015) Long-term evolution of fish communities in European mountainous rivers: past log driving effects, river management and species introduction (Salzach River, Danube). Aquat Sci 77:395–410

Hanberry BB, Kabrick JM, He HS (2015) Potential tree and soil carbon storage in a major historical floodplain forest with disrupted ecological function. Perspect Plant Ecol Evol Syst 17:17–23

Harmon ME, Franklin JF, Swanson FJ, Sollins P, Gregory SV, Lattin JD, Anderson NH, Cline SP, Aumen NG, Sedell JR, Lienkaemper GW, Cromack JR, Cummins KW (1986) Ecology of coarse woody debris in temperate ecosystems. Adv Ecol Res 15:133–302

Heimann DC, Sprague LA, Blevins DW (2011) Trends in suspended-sediment loads and concentrations in the Mississippi River basin, 1950–2009. U.S. Geological Survey Scientific Investigations Report 2011–5200, Reston, VA

Hilmes MM, Wohl EE (1995) Changes in channel morphology associated with placer mining. Phys Geogr 16:223–242

Hohensinner S, Habersack H, Jungwirth M, Zauner G (2004) Reconstruction of the characteristics of a natural alluvial river-floodplain system and hydromorphological changes following human modifications: the Danube River (1812-1991). River Res Appl 20:25–41

Holden J (2006) Sediment and particulate carbon removal by pipe erosion increase over time in blanket peatlands as a consequence of land drainage. J Geophys Res Earth 111. doi:10.1029/2005JF000386

Hood GA, Larson DG (2014) Beaver-created habitat heterogeneity influences aquatic invertebrate assemblages in boreal Canada. Wetlands 34:19–29

Hren MT, Chamberlain CP, Magilligan FJ (2001) A combined flood surface and geochemical analysis of metal fluxes in a historically mined region: a case study from the new world Mining District, Montana. Environ Geol 40:1334–1346

Hughes L (2003) Climate change and Australia: trends, projections and impacts. Austral Ecol 28:423–443

Hultine KR, Belnap J, Van Riper C, Ehleringer JR, Dennison PE, Lee ME, Nagler PL, Snyders KA, Uselman SM, West JB (2010) Tamarisk biocontrol in the western United States: ecological and societal implications. Front Ecol Environ 8:467–474

Humborg C, Ittekkot V, Cociasu A, Bodungen BV (1997) Effect of Danube River dam on Black Sea biogeochemistry and ecosystem structure. Nature 386:385–388

Imaizumi F, Sidle RC, Kamei R (2008) Effects of forest harvesting on the occurrence of landslides and debris flows in steep terrain of central Japan. Earth Surf Process Landf 33:827–840

Isaak DJ, Wollrab S, Horan D, Chandler G (2012) Climate change effects on stream and river temperatures across the northwest U.S. from 1980-2009 and implications for salmonid fisheries. Climate Change 113:499–524

Iwata T, Nakano S, Inoue M (2003) Impacts of past riparian deforestation on stream communities in a tropical rain forest in Borneo. Ecol Appl 13:461–473

Jackson CR, Martin JK, Leigh DS, West LT (2005) A southeastern piedmont watershed sediment budget: evidence for a multi-millennial agricultural legacy. J Soil Water Conserv 60:298–310

Jansson R, Nilsson C, Renöfält B (2000) Fragmentation of riparian floras in rivers with multiple dams. Ecology 81:899–903

Jiongxin X (1996) Underlying gravel layers in a large sand bed river and their influence on downstream-dam channel adjustment. Geomorphology 17:351–359

John S, Klein A (2004) Hydrogeomorphic effects of beaver dams on floodplain morphology: avulsion processes and sediment fluxes in upland valley floors (Spessart, Germany). Quaternaire 15:219–231

Johnston CA (2014) Beaver pond effects on carbon storage in soils. Geoderma 213:371–378

Kauffman JB, Krueger WC (1984) Livestock impacts on riparian ecosystems and streamside management implications … a review. J Range Manag 37:430–438

Keesstra SD, van Huissteden J, Vandenberghe J, Van Dam O, de Gier J, Pleizier ID (2005) Evolution of the morphology of the river Dragonja (SW Slovenia) due to land-use changes. Geomorphology 69:191–207

Kesel RH, Yodis EG (1992) Some effects of human modifications on sand-bed channels in southwestern Mississippi, USA. Environ Geol Water Sci 20:93–104

King J, Cambray JA, Impson ND (1998) Linked effects of dam-released floods and water temperature on spawning of the Clanwilliam yellowfish *Barbus capensis*. Hydrobiologia 384:245–265

Kleiss BA, Coupe RH, Gonthier GJ, Justus BJ (2000) Water quality in the Mississippi Embayment, Mississippi, Louisiana, Arkansas, Missouri, Tennessee, and Kentucky, 1995–98. U.S. Geological Survey Circular 1206, 36 p

Knighton AD (1989) River adjustment to changes in sediment load: the effects of tin mining on the Ringarooma River, Tasmania, 1875-1984. Earth Surf Process Landf 14:333–359

Knox JC (1977) Human impacts on Wisconsin stream channels. Ann Assoc Am Geogr 67:323–342

Knox JC (2007) The Mississippi River system. In: Gupta A (ed) Large rivers: geomorphology and management. Wiley, Chichester, UK, pp 145–182

Kondolf GM (1997) Hungry water: effects of dams and gravel mining on river channels. Environ Manag 21:533–551

Kondolf GM, Rubin ZK, Minear JT (2014) Dams on the Mekong: cumulative sediment starvation. Water Resour Res 50:5158–5169

Landwehr K, Rhoads BL (2003) Depositional response of a headwater stream to channelization, east central Illinois, USA. River Res Appl 19:77–100

Larsen MC, Parks JE (1997) How wide is a road? The association of roads and mass-wasting in a forested montane environment. Earth Surf Process Landf 22:835–848

Latocha A, Migoń P (2006) Geomorphology of medium-high mountains under changing human impact, from managed slopes to nature restoration: a study from the Sudetes, SW Poland. Earth Surf Process Landf 31:1657–1673

Lawrence S, Davies P (2014) The sludge question: the regulation of mine tailings in nineteenth-century Victoria. Environ Hist 20:385–410

Lenhardt M, Cakic P, Kolarevic J (2004) Influences of the HEPS Djerdap I and Dejerdap II dam construction on catch of economically important fish species in the Danube River. Ecohydrol Hydrobiol 4:499–502

Liébault F, Gomez B, Page M, Marden M, Peacock D, Richard D, Trotter CM (2005) Land-use change, sediment production and channel response in upland regions. River Res Appl 21L:739–756

Liermann CR, Nilsson C, Robertson J, Ng RY (2012) Implications of dam obstruction for global freshwater fish diversity. Bioscience 62:539–548

Ligon FK, Dietrich WE, Trush WJ (1995) Downstream ecological effects of dams. Bioscience 45:183–192

Liu C, Zheng H (2002) South-to-north water transfer schemes for China. Int J Water Resour Dev 18:453–471

Liu M, Tian H, Chen G, Ren W, Zhang C, Liu J (2008) Effects of land-use and land-cover change on evapotranspiration and water yield in China during 1900-2000. J Am Water Resour Assoc 44:1193–1207

Lóczy D (2007) The Danube: morphology, evolution and environmental issues. In: Gupta A (ed) Large rivers: geomorphology and management. Wiley, Chichester, UK, pp 235–260

Lucas C (2001) The Baia Mare and Baia Borsa accidents: cases of severe transboundary water pollution. Environ Policy Law 31:106–111

Luce CH, Black TA (1999) Sediment production from forest roads in western Oregon. Water Resour Res 35:2561–2570

MacDonald LH, Sampson R, Brady D, Juarros L, Martin D (2000) Predicting post-fire erosion and sedimentation risk on a landscape scale: a case study from Colorado. J Sustain For 11:57–87

Macklin MG, Brewer PA, Balteanu D, Coulthard TJ, Driga B, Howard AJ, Zaharia S (2003) The long term fate and environmental significance of contaminant metals released by the January and march 2000 mining tailings dam failures in Maramures County, upper Tisa Basin, Romania. Appl Geochem 18:241–257

Magilligan FJ, Nislow KH (2001) Long-term changes in regional hydrologic regime following impoundment in a humid-climate watershed. J Am Water Resour Assoc 37:1551–1569

Mallin MA, Johnson VL, Ensign SH (2008) Comparative impacts of stormwater runoff on water quality of an urban, a suburban, and a rural stream. Environ Monit Assess 159:475–491

May CL, Gresswell RE (2003) Processes and rates of sediment and wood accumulation in headwater streams of the Oregon coast range, USA. Earth Surf Process Landf 28:409–424

McLeay DJ, Birtwell IK, Hartman GF, Ennis GL (1987) Responses of Arctic grayling (*Thymallus arcticus*) to acute and prolonged exposure to Yukon placer mining sediment. Can J Fish Aquat Sci 44:658–673

McMahon EM, Karamanski TJ (2009) North Woods River: the St. Croix River in upper Midwest history. University of Wisconsin Press, Madison, WI

Meade RH, Moody JA (2010) Causes for the decline of suspended-sediment discharge in the Mississippi River system, 1940-2000. Hydrol Process 24:35–49

Mei-e R, Xianmo Z (1994) Anthropogenic influences on changes in the sediment load of the Yellow River, China, during the Holocene. Holocene 4:314–320

Meile T, Boillat JL, Schleiss AJ (2011) Hydropeaking indicators for characterization of the upper-Rhone River in Switzerland. Aquat Sci 73:171–182

Merritt DM, Wohl EE (2006) Plant dispersal along rivers fragmented by dams. River Res Appl 22:1–26

Miao C, Ni J, Borthwick AGL, Yang L (2011) A preliminary estimate of human and natural contributions to the changes in water discharge and sediment load in the Yellow River. Glob Planet Chang 76:196–205

Miller RR (2010) Is the past present? Historical splash dam mapping and stream disturbance detection in the Oregon Coastal Province. MS thesis, Oregon State University, Corvallis

Mirza MMQ, Warrick RA, Ericksen NJ (2003) The implications of climate change on floods of the Ganges, Brahmaputra and Meghna rivers in Bangladesh. Clim Chang 57:287–318

Morales Y, Weber LJ, Mynett AE, Newton TJ (2006) Effects of substrate and hydrodynamic conditions on the formation of mussel beds in a large river. J N Am Benthol Soc 25:664–676

Moyle PB, Mount JF (2007) Homogenous rivers, homogenous faunas. Proc Natl Acad Sci U S A 104:5711–5712

Muhar S, Jungwirth M, Unfer G, Wiesner C, Poppe M, Schmutz S, Hohensinner S, Habersack H (2008) Restoring riverine landscapes at the Drau River: successes and deficits in the context of ecological integrity. In: Habersack H, Piégay H, Rinaldi M (eds) Gravel-bed rivers VI: from process understanding to river restoration. Elsevier, Amsterdam, pp 703–737

Musselman ZA (2011) The localized role of base level lowering on channel adjustment of tributary streams in the Trinity River downstream of Livingston dam, Texas, USA. Geomorphology 128:42–56

Myers TJ, Swanson S (1996) Long-term aquatic habitat restoration: Mahogany Creek, Nevada, as a case study. Water Resour Bull 32:241–252

Nachtnebel H-P (2000) The Danube River basin environmental programme: plans and actions for a basin wide approach. Water Policy 2:113–129

Naiman RJ, Pinay G, Johnston CA, Pastor J (1994) Beaver influences on the long-term biogeochemical characteristics of boreal forest drainage networks. Ecology 75:905–921

Naiman RJ, Decamps H, McClain ME (2005) Riparia: ecology, conservation, and management of streamside communities. Elsevier, Amsterdam

Nik AR (1988) Water yield changes after forest conversion to agricultural land use in peninsular Malaysia. J Trop For Sci 1:67–84

Nilsson C, Berggren K (2000) Alterations of riparian ecosystems caused by river regulation. Bioscience 50:783–792

Nilsson C, Lepori F, Malmqvist B, Törnlund E, Hjerdt N, Helfield JM, Palm D, Östergren J, Jansson R, Brännäs E, Lundqvist H (2005a) Forecasting environmental responses to restoration of rivers used as log floatways: an interdisciplinary challenge. Ecosystems 8:779–800

Nilsson C, Reidy CA, Dynesius M, Revenga C (2005b) Fragmentation and flow regulation of the world's large river systems. Science 308:405–408

Nilsson C, Polvi LE, Lind L (2015) Extreme events in streams and rivers in arctic and subarctic regions in an uncertain future. Freshw Biol 60:2535–2546

Nolet BA, Rosell F (1998) Comeback of the beaver *Castor fiber*: an overview of old and new conservation problems. Conserv Biol 83:165–173

Ogden FL, Pradhan NR, Downer CW, Zahner JA (2011) Relative importance of impervious area, drainage density, width function, and subsurface storm drainage on flood runoff from an urbanized catchment. Water Resour Res 47. doi:10.1029/2011WR010550

Ortlepp J, Mürle U (2003) Effects of experimental flooding on brown trout (Salmo Trutta fario L.): the river Spöl, Swiss National Park. Aquat Sci 65:232–238

Palmer MA, Lettenmaier DP, Poff NL, Postel SL, Richter B, Warner R (2009) Climate change and river ecosystems: protection and adaptation options. Environ Manag 44:1053–1068

Palmer MA, Bernhardt ES, Schlesinger WH, Eshleman KN, Foufoula-Georgiou E, Hendryx MS, Lemly AD, Likens GE, Loucks OL, Power ME, White PS, Wilcock PR (2010) Mountaintop mining consequences. Science 327:148–149

Pastur GM, Lencinas MV, Escobar J, Quiroga P, Malmierca L, Lizarralde M (2006) Understorey succession in Nothofagus forests in Tierra del Fuego (Argentina) affected by *Castor canadensis*. Appl Veg Sci 9:143–154

Patten DT (1998) Riparian ecosystems of semi-arid North America: diversity and human impacts. Wetlands 18:498–512

Patten DT, Harpman DA, Voita MI, Randle TJ (2001) A managed flood on the Colorado River: background, objectives, design, and implementation. Ecol Appl 11:635–643

Paxton EH, Theimer TC, Sogge MK (2011) Tamarisk biocontrol using tamarisk beetles: potential consequences for riparian birds in the southwestern United States. Condor 113:255–265

Peipoch M, Brauns M, Hauer RF, Weitere M, Valett HM (2015) Ecological simplification: human influences on riverscape complexity. Bioscience 65:1057–1065

Peng J, Chen S, Dong P (2010) Temporal variation of sediment load in the Yellow River basin, China, and its impacts on the lower reaches and the river delta. Catena 83:135–147

Perna C, Burrows D (2005) Improved dissolved oxygen status following removal of exotic weed mats in important fish habitat lagoons of the tropical Burdekin River floodplain, Australia. Mar Pollut Bull 51:138–148

Pierce JL, Meyer GA, Jull AJT (2004) Fire-induced erosion and millennial-scale climate change in northern ponderosa pine forests. Nature 432:87–90

Pinter N (2005) One step forward, two steps back on U.S. floodplains. Science 308:207–208

Pitt F, Field R, Lalor M, Brown M (1995) Urban stormwater toxic pollutants: assessment, sources and treatability. Water Environ Res 67:260–275

Pizzuto J (2002) Effects of dam removal on river form and process. Bioscience 52:683–691

Poff NL, Hart DD (2002) How dams vary and why it matters for the emerging science of dam removal. Bioscience 52:659–668

Poff NL, Olden JD, Merritt DM, Pepin DM (2007) Homogenization of regional river dynamics and global biodiversity implications. Proc Natl Acad Sci U S A 104:5732–5737

Pollock MM, Heim M, Werner D (2003) Hydrologic and geomorphic effects of beaver dams and their influence on fishes. In: Gregory SV, Boyer K, Gurnell A (eds) The ecology and management of wood in world rivers, American Fisheries Society Symposium, vol 37, pp 213–233

Pollock MM, Beechie TJ, Jordan CE (2007) Geomorphic changes upstream of beaver dams in Bridge Creek, an incised stream channel in the interior Columbia River basin, eastern Oregon. Earth Surf Process Landf 32:1174–1185

Pollock MM, Beechie TJ, Wheaton JM, Jordan CE, Bouwes N, Weber N, Volk C (2014) Using beaver dams to restore incised stream ecosystems. Bioscience 64:279–290

Polvi LE, Wohl E (2012) The beaver meadow complex revisited—the role of beavers in post-glacial floodplain development. Earth Surf Process Landf 37:332–346

Polvi LE, Wohl E (2014) Biotic drivers of stream planform: implications for understanding the past and restoring the future. Bioscience 63:439–452

Pringle CM, Freeman MC, Freeman BJ (2000) Regional effects of hydrologic alterations on riverine macrobiota in the new world: tropical-temperate comparisons. Bioscience 50:807–823

Rabalais NN, Turner RE (2001) Hypoxia in the northern Gulf of Mexico: description, causes and change. In: Rabalais NN, Turner RE (eds) Coastal hypoxia: consequences for living resources and ecosystems. Coastal and Estuarine Studies 58. American Geophysical Union Press, Washington, DC, pp 1–36

Rabalais NN, Turner RE, Sen Gupta BK, Boesch DF, Chapman P, Murrell MC (2007) Hypoxia in the northern Gulf of Mexico: does the science support the plan to reduce, mitigate, and control hypoxia? Estuar Coasts 30:753–772

Renshaw CE, Abengoza K, Magilligan FJ, Dade WB, Landis JD (2014) Impact of flow regulation on near-channel floodplain sedimentation. Geomorphology 205:120–127

Rhoads BL (1990) The impact of stream channelization on the geomorphic stability of an arid-region river. Natl Geogr Res 6:157–177

Ripple WJ, Beschta RL (2004) Wolves and the ecology of fear: can predation risk structure ecosystems? Bioscience 54:755–766

Roberts CR (1989) Flood frequency and urban-induced channel change: some British examples. In: Beven K, Carling P (eds) Floods: hydrological, sedimentological and geomorphological implications. Wiley, Chichester, UK, pp 57–82

Rood SB, Pan J, Gill KM, Franks CG, Samuelson GM, Shepherd A (2008) Declining summer flows of Rocky Mountain rivers: changing seasonal hydrology and probable impacts on floodplain forests. J Hydrol 349:397–410

Rooney RC, Bayley SE, Schindler DW (2012) Oil sands mining and reclamation cause massive loss of peatland and stored carbon. Proc Natl Acad Sci U S A 109:4933–4937

Rosell F, Bozser O, Collen P, Parker H (2005) Ecological impact of beavers *Castor fiber* and *Castor canadensis* and their ability to modify ecosystems. Mammal Rev 35:248–276

Ryan S (1997) Morphologic response of subalpine streams to transbasin flow diversion. J Am Water Resour Assoc 33:839–854

Sabo JL, Sponseller R, Dixon M, Gade K, Harms T, Heffernan J, Jani A, Katz G, Soykan C, Watts J, Welter J (2005) Riparian zones increase regional species richness by harboring different, not more, species. Ecology 86:56–62

Salant NL, Renshaw CE, Magilligan FJ (2006) Short and long-term changes to bed mobility and bed composition under altered sediment regimes. Geomorphology 76:43–53

Scarnecchia DL (1988) The importance of streamlining in influencing fish community structure in channelized and unchannelized reaches of a prairie stream. Regul Rivers Res Manag 2:155–166

Schama S (1995) Landscape and memory. Knopf, New York, NY

Schmocker-Fackel P, Naef F (2010) More frequent flooding? Changes in flood frequency in Switzerland since 1850. J Hydrol 381:1–8

Schoof R (1980) Environmental impact of channel modification. Water Resour Bull 16:697–701

Schreck P (1998) Environmental impact of uncontrolled waste disposal in mining and industrial areas in central Germany. Environ Geol 35:66–72

Schulze DJ, Walker KF (1997) Riparian eucalypts and willows and their significance for aquatic invertebrates in the river Murray, South Australia. Regul Rivers Res Manag 13:557–577

Sedell JR, Froggatt JL (1984) Importance of streamside forests to large rivers: the isolation of the Willamette River, Oregon, USA, from its floodplain by snagging and streamside forest removal. Verh Int Verein Limnol 22:1828–1834

Sedell JR, Leone FN, Duval WS (1991) Water transportation and storage of logs. In: Meehan WR (ed) Influences of forest and rangeland management on salmonid fishes and their habitats, American Fisheries Society Symposium, vol 19, pp 325–368

Shafroth PB, Cleverly JL, Dudley TL, Taylor JP, Van Riper C, Weeks EP, Stuart JN (2005) Control of *Tamarix* in the western United States: implications for water salvage, wildlife use, and riparian restoration. Environ Manag 35:231–246

Shafroth PB, Wilcox AC, Lytle DA, Hickey JT, Andersen DC, Beauchamp VB, Hautzinger A, McMullen LE, Warner A (2010) Ecosystem effects of environmental flows: modelling and experimental floods in a dryland river. Freshw Biol 55:68–85

Shearer KD, Mulley JC (1978) The introduction and distribution of the carp, *Cyprinus carpio* Linnaeus, in Australia. Aust J Mar Freshwat Res 29:551–563

Shen HW (1999) Flushing sediment through reservoirs. J Hydraul Res 37:743–757

Simco AH, Stephens DB, Calhoun K, Stephens DA (2009) Historic irrigation and drainage at Priestley farm by Joseph Elkington and William smith. Vadose Zone J 9:4–13

Skalak KJ, Benthem AJ, Schenk ER, Hupp CR, Galloway JM, Nustad RA, Wiche GJ (2013) Large dams and alluvial rivers in the Anthropocene: the impacts of the garrison and Oahe dams on the upper Missouri River. Anthropocene 2:51–64

Smith N (1971) A history of dams. Peter Davies, London

Spaling H, Smit B (1995) A conceptual model of cumulative environmental effects of agricultural land drainage. Agric Ecosyst Environ 53:99–108

Stanford JA, Hauer FR (1992) Mitigating the impacts of stream and lake regulation in the Flathead River catchment, Montana, USA: an ecosystem perspective. Aquat Conserv Mar Freshw Ecosyst 2:35–63

Stanley EH, Doyle MW (2003) Trading off: the ecological effects of dam removal. Front Ecol Environ 1:15–22

Stewart IT, Cayan DR, Dettinger MD (2005) Changes toward earlier streamflow timing across western North America. J Clim 18:1136–1155

Stinchcomb GE, Messner TC, Driese SG, Nordt LC, Stewart RM (2011) Pre-colonial (a.D. 1100-1600) sedimentation related to prehistoric maize agriculture and climate change in eastern North America. Geology 39:363–366

Stoughton JA, Marcus WA (2000) Persistent impacts of trace metals from mining on floodplain grass communities along soda Butte Creek, Yellowstone National Park. Environ Manag 25:305–320

Strayer DL, Caraco NF, Cole JJ, Findlay S, Pace ML (1999) Transformation of freshwater ecosystems by bivalves: a case study of zebra mussels in the Hudson River. Bioscience 49:19–27

Strayer DL, Downing JA, Haag WR, King TL, Layzer JB, Newton TJ, Nichols JS (2004) Changing perspectives on pearly mussels, North America's most imperiled animals. Bioscience 54:429–439

Sumi T, Kantoush SA (2010) Integrated management of reservoir sediment routing by flushing, replenishing, and bypassing sediments in Japanese river basins. In: Proceedings of the 8th International Symposium on Ecohydrology, Kyoto, Japan, pp 831–838

Surian N (1999) Channel changes due to river regulation: the case of the Piave River, Italy. Earth Surf Process Landf 24:1135–1151

Sutter PS (2015) Let us now praise famous gullies: providence canyon and the soils of the south. University of Georgia Press, Atlanta, GA

Swanson BJ, Meyer GA, Coonrod JE (2011) Historical channel narrowing along the Rio Grande near Albuquerque, New Mexico in response to peak discharge reductions and engineering: magnitude and uncertainty of change from air photo measurements. Earth Surf Process Landf 36:885–900

Ta W, Xiao H, Dong Z (2008) Long-term morphodynamic changes of a desert reach of the Yellow River following upstream large reservoirs' operation. Geomorphology 97:249–259

Tobin GA (1995) The levee love affair: a stormy relationship? J Am Water Resour Assoc 31:359–367

Tockner K, Schiemer F, Ward JV (1998) Conservation by restoration: the management concept for a river-floodplain system on the Danube River in Austria. Aquat Conserv Mar Freshw Ecosyst 8:71–86

Törnlund E, Östlund L (2002) Floating timber in northern Sweden: the construction of floatways and transformation of rivers. Environ Hist 8:85–106

Townsend CR, Crowl TA (1991) Fragmented population structure in New Zealand fish: an effect of introduced brown trout? Oikos 61:347–354

Trimble SW, Mendel AC (1995) The cow as a geomorphic agent—a critical review. Geomorphology 13:233–253

Turner RE (1997) Wetland loss in the northern Gulf of Mexico: multiple working hypotheses. Estuaries 20:1–13

Ulloa H, Iroumé A, Lenzi MA, Andreoli A, Álvarez C, Barrera V (2011) Large wood in two catchments from the Coastal Mountain range with different land use history. Bosque 32:235–245. [Spanish with English abstract]

US FWS (US Fish and Wildlife Service) (2015) The Beaver Restoration Guidebook, v. 1.0. Portland, OR, 189 pp. http://www.fws.gov/oregonfwo/ToolsForLandowners/RiverScience/Beaver.asp

Van Nieuwenhuyse EE, LaPerriere JD (1986) Effects of placer gold mining on primary production in subarctic streams of Alaska. Water Resour Bull 22:91–99

Vietz GJ, Sammonds MJ, Stewardson MJ (2013) Impacts of flow regulation on slackwaters in river channels. Water Resour Res 49:1797–1811

Vitousek PM, Naylor R, Crews T, David MB, Drinkwater LE, Holland E, Johnes PJ, Katzenberger J, Martinelli LA et al (2009) Nutrient imbalances in agricultural development. Science 324:1519–1520

Walker KF, Thoms MC (1993) Environmental effects of flow regulation on the lower Murray River, Australia. Regul Rivers Res Manag 8:103–119

Wang ZY, Wu B, Wang G (2007a) Fluvial processes and morphological response in the yellow and Weihe rivers to closure and operation of Sanmenxia dam. Geomorphology 91:65–79

Wang H, Yang Z, Saito Y, Liu JP, Sun X, Wang Y (2007b) Stepwise decreases of the Huanghe (Yellow River) sediment load (1950-2005): impacts of climate change and human activities. Glob Planet Chang 57:331–354

Wang M, Qi S, Zhang X (2012) Wetland loss and degradation in the Yellow River Delta, Shandong Province of China. Environ Earth Sci 67:185–188

Ward JV (1998) Riverine landscapes: biodiversity patterns, disturbance regimes, and aquatic conservation. Biol Conserv 83:269–278

Wegener P, Covino T, Wohl E (2017) Beaver-mediated lateral hydrologic connectivity, fluvial carbon and nutrient flux, and aquatic ecosystem metabolism. Water Resour Res 53:4606–4623

Wenger SJ, Isaak DJ, Luce CH, Neville HM, Fausch KD, Dunham JB, Dauwalter DC, Young MK, Elsner MM, Rieman BE, Hamlet AF, Williams JE (2011) Flow regime, temperature, and biotic interactions drive differential declines of trout species under climate change. Proc Natl Acad Sci U S A 108:14175–14180

Wickham J, Wood PB, Nicholson MC, Jenkins W, Druckenbrod D, Suter GW, Strager MP, Mazzarella C, Galloway W, Amos J (2013) The overlooked terrestrial impacts of mountaintop mining. Bioscience 63:335–348

Wilcox AC, Shafroth PB (2013) Coupled hydrogeomorphic and woody-seedling response to controlled flood releases in a dryland river. Water Resour Res 49:2843–2860

Williams GP, Wolman MG (1984) Downstream effects of dams in alluvial rivers. U.S. Geological Survey Professional Paper 1286. US Government Printing Office, Washington, DC

Wiseman WJ, Rabalais NN, Turner RE, Dinnel SP, MacNaughton A (1997) Seasonal and interannual variability within the Louisiana coastal current: stratification and hypoxia. J Mar Syst 12:237–248

Wohl E (2004) Disconnected rivers: linking rivers to landscapes. Yale University Press, New Haven, CT

Wohl E (2011) A world of rivers: environmental change on ten of the World's great rivers. University of Chicago Press, Chicago, IL

Wohl E (2013) Wide rivers crossed: the South Platte and the Illinois of the American prairie. University Press of Colorado, Boulder, CO

Wohl E (2014) A legacy of absence: wood removal in U.S. Rivers. Prog Phys Geogr 38:637–663

Wohl E, Dust D (2012) Geomorphic response of a headwater channel to augmented flow. Geomorphology 138:329–338

Wolman MG (1967) A cycle of sedimentation and erosion in urban river channels. Geogr Ann 49A:385–395

Wright JP (2009) Linking populations to landscapes: richness scenarios resulting from changes in the dynamics of an ecosystem engineer. Ecology 90:3418–3429

WWF (World Wildlife Fund) (2013) Freshwater ecoregions of the world. http://www.feow.org/index.php

Wyzga B (2001) A geomorphologist's criticism of the engineering approach to channelization of gravel-bed rivers: case study of the Raba River, polish Carpathians. Environ Manag 28:341–358

Young MK, Haire D, Bozek MA (1994) The effect and extent of railroad tie drives in streams of southeastern Wyoming. West J Appl For 9:125–130

Yu L (2002) The Huanghe (yellow) river: a review of its development, characteristics, and future management issues. Cont Shelf Res 22:389–403

Chapter 4
Toward Sustainable Rivers and Water Resources

The previous chapter reviewed a long list of human activities that have directly and indirectly altered process and form in river ecosystems, with an associated loss of ecosystem services. Management in many river basins now focuses on trying to restore some balance between existing, typically consumptive or homogenizing uses of river resources versus restoration of diversity, sustainability, and river ecosystem health. None of the latter three terms is easy to define. Diversity refers to variety but, as discussed in Chap. 2, there are multiple ways to define physical or biotic diversity and no particular metric is consistently better in all situations or for all purposes. Sustainability in an ecological context typically refers to the ability of ecosystems to remain diverse and productive. In a physical context, sustainability may refer to the ability of a system to continue functioning or providing natural resources. River health also has multiple potential definitions. Even using the relatively simple definition in Chap. 1, that river health is the degree to which a river ecosystem matches natural conditions, requires understanding natural conditions. On the one hand, use and understanding of words matters and people may use similar words but have different objectives or desired outcomes. On the other hand, diversity, sustainability, and river health are now widely used and there is at least broad consensus on the meaning of these words. So, what is being done to protect and restore river corridors? This chapter reviews the development of river restoration during the latter twentieth century and early twenty-first century; discusses particular concepts used in restoration; and presents specific examples of watershed-scale restoration.

4.1 River Restoration

River restoration is a form of river management designed to assist the establishment of improved hydrologic, geomorphic, and ecological processes in a degraded watershed and to replace lost, damaged, or compromised elements of the natural system

© The Author(s) 2018
E. Wohl, *Sustaining River Ecosystems and Water Resources*, SpringerBriefs in Environmental Science, DOI 10.1007/978-3-319-65124-8_4

(Wohl et al. 2005). This definition is broad in that it allows for diverse interpretations of what constitutes improved. Improvements could include activities as diverse as stabilizing banks to protect private property or introducing large wood to enhance fish habitat (Abbe and Brooks 2011). Ecological river restoration can be further specified as restoration that assists the recovery of ecological integrity in a degraded watershed by reestablishing the processes necessary to support the natural river ecosystem within the watershed (Wohl et al. 2005). Restoration is sometimes distinguished from rehabilitation by emphasizing that restoration focuses on processes that create and maintain a desired river condition, whereas rehabilitation is more likely to focus on form, such as stabilizing stream banks or beds. In this chapter, restoration is used in a broad, generic sense and includes rehabilitation.

Restoration designed to increase the esthetic appeal of a river has been undertaken for centuries. Kondolf (2006) traces the still-widespread preference for a slightly sinuous channel bordered by open, park-like woodland to the parks designed for wealthy British landowners by late eighteenth-century landscape architect Capability Brown (Fig. 4.1). Restoration designed to enhance specific river functions also goes back more than century, primarily in the context of recreational fishing, as traced in detail by Douglas Thompson in his book *The Quest for the Golden Trout* (Thompson 2013).

Restoration as a multibillion-dollar industry, however, is a phenomenon of the very late twentieth and early twenty-first centuries (Bernhardt et al. 2005). The objective of many restoration projects has shifted during this period. Many of the first restoration projects focused on creating a desired channel configuration at the reach-scale of tens to hundreds or thousands of meters in length, typically to enhance fisheries, esthetic appeal, or to improve water quality. Recently, restoration has focused more on creating desired processes, commonly for similar endpoints of fisheries or water quality enhancement (Beechie et al. 2010; Wohl et al. 2015b). Desired processes include channel-floodplain connectivity (Tockner et al. 1999; Hughes et al. 2001; Shields et al. 2011; Gumiero et al. 2013), hyporheic exchange (Hester and Gooseff 2011), ecological productivity (Lepori et al. 2005; Palmer et al. 2010a, b), longitudinal connectivity (Shafroth et al. 2010; Konrad et al. 2011), and the space to adjust to changes in water and sediment inputs (Kondolf 2011).

Bernhardt and Palmer (2011) distinguish reconfiguration and reconnection. Reconfiguration efforts are designed to change the physical structure of the river corridor through reshaping, replanting, or reconstruction. Reconnection efforts involve the removal or retrofitting of infrastructure that was previously installed to limit connectivity within the river corridor, such as dams and levees. Although reconnection or reconfiguration can be undertaken at any spatial scale, reach-scale projects are more likely to focus on reconfiguration and watershed-scale projects on reconnection (Wohl et al. 2015b). Kondolf et al. (2006) develop simple conceptual models and bivariate plots to visualize how human alterations have changed different dimensions of connectivity and how restoration can mitigate these changes.

The majority of river restoration projects are undertaken in the United States (Bernhardt et al. 2005, 2007), Europe and the UK (Brookes 1990; Sear 1994; McDonald et al. 2004; Clifford 2012), and Australia (Brierley and Fryirs 2000,

Fig. 4.1 Opposing illustrations of river health from the inside front cover of the textbook River Pollution (Haslam 1994)

2005; Brooks et al. 2006; Lester and Boulton 2008), but restoration is becoming more common in other regions (e.g., Yaning et al. 2006 for China; Nakamura et al. 2006 for Japan).

Although there is substantial overlap, different restoration approaches are emphasized in small to medium-sized rivers versus medium-sized to large rivers. Small, steep rivers can be targets for restoration in mountainous regions with a history of bank stabilization and grade-control structures (Fig. 4.2). Restoration in these channels may focus on replacing grade-control or check dams with artificial steps designed using natural materials and morphological characteristics of natural step-pool sequences (Chin et al. 2009; Comiti et al. 2009).

Small rivers in urban environments, mining regions, or agricultural regions are also targets for restoration. Infrastructure and private property severely constrain restoration options in most urban areas. However, enhancement of vegetated infiltration zones, acquisition of at least limited floodplain areas, and replanting of riparian vegetation can create urban amenities, improve water quality, and reduce flood hazards (Palmer et al. 2014).

Areas impacted by mining can have completely altered topography, large sediment fluxes, reduced capacity for vegetative colonization, and chemical contamination, all of which can limit restoration of biogeochemical and ecological functions

Fig. 4.2 Grade-control structures in the Alptal of Switzerland (*upper photo*; channel is approximately 15 m wide) and bank stabilization along a tributary of the Schwarzbach in the town of Sankt Jakob in Defereggen, Austria (*lower photo*; channel is approximately 6 m wide)

of rives (Palmer and Hondula 2014). Much more effort needs to be devoted to understanding how to restore process and form in these river corridors.

Small rivers in agricultural areas have been channelized (Fig. 4.3), buried in pipes (Weissmann et al. 2009), and subjected to excess nutrients and other

Fig. 4.3 Downstream view of a channelized portion of Partridge Creek, a tributary of the Illinois River, USA, that is now being allowed to develop some irregularities of channel cross-sectional geometry. The channel is about 10 m wide and drains about 75 km² here at 40.851°N, 89.464°W

agricultural chemicals in runoff. Restoration of channelized streams includes reme-andering (Wade et al. 2002; Lorenz et al. 2009) by artificially creating a sinuous channel that is then allowed to adjust to some degree. Restoration can also involve cessation of dredging and bank stabilization (Rhoads and Herricks 1996). These activities can result in limited restoration of ecological function, but continued inputs of contaminants may limit the ecological benefits of reconfiguration activi-ties (Roley et al. 2012). Reduction or elimination of riparian grazing in agricultural areas has been a particularly successful approach to restoring diverse riparian veg-etation, stream bank stability, water quality, and aquatic communities (Rhodes et al. 2007; Hickford and Schiel 2014). Reintroduction of wood in agricultural streams with historically forested catchments can enhance macroinvertebrate and fish diver-sity, increase sediment and organic matter storage, and improve bed and bank stabil-ity (Lester and Boulton 2008).

Restoration on medium-sized to large rivers can involve reconfiguration that cre-ates reconnection, such as levee setbacks or notching (Florsheim and Mount 2002; Hughes and Rood 2003; Zhang and Mitsch 2007; Konrad et al. 2008; Marks et al. 2014; Nakamura et al. 2014). Removal of structures that block access to secondary channels or creating structures to block flow from channelized sections and redirect flow toward naturally created meanders (Koebel and Bousquin 2014) can also reconnect river corridors. Removing dams (Major et al. 2012; Wilcox et al. 2014; East et al. 2015; O'Connor et al. 2015) or changing water and sediment releases from the dam (Galat et al. 1998; Ortlepp and Murle 2003; Konrad et al. 2011; Melis 2011; Flessa et al. 2013; Mueller et al. 2014) can strongly increase longitudinal connectivity, as well as lateral and vertical connectivity, by creating more natural flow regimes.

An important consideration is that restoration is undertaken in a human or societal context in which perceptions of desired river characteristics and river health are of critical importance. Reach-scale restoration, in particular, is more likely to be driven, or at least strongly influenced, by stakeholders in local communities, whereas watershed-scale restoration is more commonly driven by government resource-management agencies. Reach-scale restoration may involve esthetic or recreational enhancements that do not necessarily improve ecological functions in a river corridor (Bernhardt et al. 2007). Reach-scale restoration sometimes also seeks to create conditions that are not natural or ecologically appropriate for a particular river corridor (Brierley and Fryirs 2009; Fryirs and Brierley 2009), as in the many examples of braided rivers that are viewed as esthetically unattractive and reconfigured to a meandering planform (Kondolf et al. 2001). These scenarios can occur where there is a gap between scientific understanding of a river ecosystem and societal perceptions.

Large wood in river corridors illustrates this gap very well. An extensive literature documents the numerous beneficial physical and ecological effects of wood in rivers (Gregory et al. 2003; Wohl 2017) and wood was abundant historically in forested river corridors in all regions of the world (Gregory et al. 2003; Wohl 2014). Nonetheless, much of contemporary river management and restoration continues to remove wood from channels and floodplains either because of perceived safety hazards or simply for esthetic reasons (Wohl 2015). Surveys conducted among university students in introductory geology or physical geography classes in the United States and Europe indicate that most students have highly negative perceptions of wood in river corridors (Chin et al. 2008), whereas similar surveys indicate generally positive perceptions of wood among experienced river managers (Chin et al. 2014).

After decades of river restoration projects, how successful have these efforts been? Numerous evaluations of river restoration by river scientists emphasize three points (Wohl et al. 2015b). First, most restoration projects, especially reconfiguration projects on small to medium-sized rivers, do not include monitoring to quantitatively and objectively evaluate whether restoration goals were achieved (e.g., Bernhardt et al. 2005). Souchon et al. (2008), for example, note that there has been little validation monitoring of the response of aquatic species to restored flow regimes. Second, a substantial proportion of restoration projects do not achieve significant improvements in river function as reflected in water quality, ecological communities, or ecosystem services (Lepori et al. 2005; Bernhardt and Palmer 2011; Violin et al. 2011; Palmer and Hondula 2014). Third, river scientists and managers must work to more effectively involve the non-scientific community in planning and implementation of river restoration (Eden et al. 2000; Pfadenhauer 2001; Wade et al. 2002; Eden and Tunstall 2006; Eden and Bear 2011). There are numerous explanations for why many restoration projects do not achieve significant improvements. These include restoration activities that are not appropriate for restoring processes or that are hampered by continuing limitations on flooding associated with flow regulation (Kristensen et al. 2014). In some projects, the small-scale effects of appropriate restoration do not significantly ameliorate the continued, watershed-scale effects of stressors such as introduced species or excess nutrients and other contaminants entering the river corridor (Bernhardt and Palmer 2011). This does not mean, however, that river restoration is not necessary and vitally important. It means that we must learn how to more effectively restore river corridors and how to measure the effectiveness of restoration.

Palmer et al. (2005) propose five criteria for assessing the ecological success of river restoration: (1) Restoration should be designed around a specific guiding image of a dynamic, healthy river that could exist at the site targeted for restoration. 'Could exist' is a critical phrase in this context. Although a meandering channel may be considered desirable, temporal fluctuations in water and sediment inputs may be more characteristic of a braided channel planform, for example. The key point is to understand the context of the river or river segment targeted for restoration and its trajectories of change through time (Campana et al. 2014; Petkovska et al. 2015; Brierley and Fryirs 2016). (2) The river corridor's ecological condition must be measurably improved, as assessed by movement toward the guiding image in terms of parameters such as water quality, improved bioabundance or biodiversity. This criterion is designed to counteract the tendency to evaluate river restoration solely on visual appeal or physical characteristics. This criterion also incorporates interactions between process and form. Restoration of substrate heterogeneity in headwater streams of Finland, for example, did not result in full restoration of retention efficiency because the aquatic mosses critical to retention were removed during restoration activities (Muotka and Laasonen 2002). (3) The river ecosystem must be more self-sustaining and resilient to disturbances. A river restoration project that requires continual maintenance after restoration indicates that the restored channel is not effectively adjusted to temporal fluctuations in water and sediment inputs. (4) No lasting harm should be done to the river corridor during the construction phase of restoration. (5) Pre- and post-assessment must be completed and data made publicly available (Palmer et al. 2007).

Because water and sediment inputs are among the primary drivers of river process and form, river management and restoration over the past few decades have emphasized flow regime. The most recent expression of this emphasis is the concept of environmental flows.

4.2   Environmental Flows

Initial efforts to protect flows in river ecosystems focused on minimum flows. In drylands such as the western United States, consumptive flow uses can create river segments that are completely dewatered continuously or for discrete periods of time. Fish biologists developed the concept of instream flows as a mechanism for preserving a minimum flow within the active channel in order to sustain fish populations. Early approaches used a biological model of fish habitat preference by species. The biological model quantified optimal flow depth, velocity, and substrate type. The biological model was combined with a hydraulic model of changes in habitat availability based on discharge (Bovee and Milhous 1978). Although this approach does not account for other limitations such as biological competition or predation or limited physical connectivity, this type of model remains widely used in evaluating alternative water management scenarios (Stalnaker et al. 1995; Grigg 2016).

Limitations associated with focusing only on minimum flows led to the idea of protecting channel maintenance flows. Channel maintenance flows are components of a river flow regime necessary to maintain specific channel characteristics, such as cross-sectional area for flood conveyance or substrate grain-size distribution for fish spawning. The concept of channel maintenance flows explicitly recognizes that discharge variability is necessary to sustain physical and biotic components of a river ecosystem and that this variability can be characterized in terms of thresholds of flow magnitude and duration necessary to maintain river processes and forms. Channel maintenance flows can focus on a limited objective, such as pool scour, or incorporate a broader range of flow magnitudes designed to maintain a physically diverse river corridor (Andrews and Nankervis 1995).

Channel maintenance flows designed to maintain multiple aspects of river form gave rise to the current focus on environmental flows. Environmental flows were initially experimental flow releases from dams. These flows were designed to restore specific aspects of the downstream river ecosystem by mimicking naturally occurring floods, although the experimental releases are typically of lower magnitude and shorter duration than natural floods (Galat et al. 1998; Mürle et al. 2003; Konrad et al. 2011; Melis 2011; Flessa et al. 2013). Description of the natural flow regime of a river (Poff et al. 1997) broadened environmental flows from experimental flood releases to quantifying an annual hydrograph that specifies magnitude, frequency, timing, duration, and rate of change in flow. Assessing environmental flow requirements typically involves quantifying natural and altered stream flows and changes in the flow regime (Richter et al. 1996; Gao et al. 2009), as well as quantifying relationships between hydrologic metrics and physical and biotic river attributes (Sanderson et al. 2012). Although both of these steps can be challenging, environmental flow assessments are now widely used in a variety of contexts (e.g., Tharme 2003; Arthington et al. 2006; Rathburn et al. 2009; Poff and Zimmerman 2010; Shafroth et al. 2010; Kozak et al. 2016).

The underlying intent of environmental flows is to reach a compromise between consumptive water use that takes the entire river flow versus a completely natural flow regime. This compromise recognizes the need to maintain physical and biotic characteristics of a river ecosystem while maintaining water supplies for consumptive use (Fig. 4.4).

The need to protect endangered species drives many environmental flows. Ecologists tend to focus on flow regime, partly because of the enormous influence

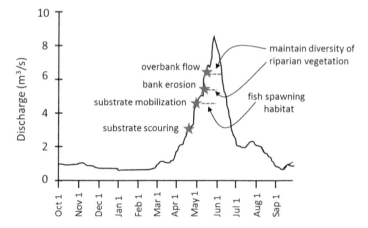

Fig. 4.4 Hypothetical illustration of environmental flow parameters using an average annual snowmelt hydrograph. Magnitude thresholds indicated by *red stars*. The lowest star indicates initiation of scour of stable substrate, which can remove periphyton from bed sediment and affect nutrient spiraling and macroinvertebrate habitat. Substrate mobilization refers to initiation of motion for bed sediment, which can flush silt and clay that clogs interstices between coarser bed grains and limits suitability of the bed for fish spawning habitat. Erosion of stream banks can remove senescent riparian vegetation and provide germination sites for new plants after the flow recedes. Overbank flow can sustain riparian vegetation, provide floodplain spawning or nursery sites for fish, and create a flux of nutrients back into the channel as flows decline. Duration thresholds indicated by *dashed red lines* illustrate minimum duration of a specified flow magnitude necessary to attain desired ecosystem processes. Ideally, an environmental flow regime would include flows that meet or exceed each of these magnitude and duration thresholds, although such flows might not occur every year

of the natural flow regime paper of Poff et al. (1997). Geomorphologists increasingly emphasize the equal importance of sediment dynamics in maintaining physically complex and connected river ecosystems (Pitlick and Wilcock 2001; Schmidt and Wilcock 2008; Wohl et al. 2015a), not least because altered flow regimes typically correspond to altered sediment dynamics as a result of changes in sediment supply and transport capacity. Sediment dynamics in the context of environmental flows is now increasingly the target of river research and management (Rubin et al. 1998; Wiele et al. 2007; Klösch et al. 2011).

Geomorphologists are now also starting to emphasize the importance of large wood dynamics in forested regions (Kail et al. 2007) because large wood inputs interact with water and sediment inputs to strongly influence river process and form (Gurnell et al. 2005). Large wood also provides an important nutrient source in channels and floodplains (Pettit and Naiman 2006). The emphasis on large wood is a much more recent development in river geomorphology, partly because of the relatively recent appreciation for the enormous historical reductions in wood recruitment to river corridors and wood loads within river corridors (Collins et al. 2002). Having become accustomed to wood-poor river corridors, river scientists have been slow to recognize the vital role played by wood in forested river ecosystems.

4.3   Integrative Planning

One component of more effective river restoration is thinking carefully about integration across space and through time: that is the intention of the phrase 'integrative planning.' Integration across diverse spatial scales is critically important because of the many forms of connectivity within a watershed and between the watershed and the greater environment. Many forms of river restoration are implemented within a limited spatial scale: the operating regime of a single dam is modified or the erosional resistance of a limited length of river bank is enhanced by restoring woody riparian vegetation. The ability of restoration activities to achieve desired ends, however, may well depend on factors outside the restoration area. Modifying the operating regime of the dam may not restore desired native fish communities, for example, because introduced species continue to outcompete native species. Recovery of woody riparian vegetation may not sufficiently increase nitrate uptake and processing because nitrate loads coming from upstream portions of the watershed overwhelm the uptake capacity of the restored segment of river corridor.

Thinking about a segment of river corridor as connected longitudinally, laterally, and vertically also facilitates consideration of factors outside of the active channel that might limit restoration success. An example comes from a project to restore native riparian willows along a formerly vegetated river corridor in California, USA. Willows planted at the site were unable to survive because ground water pumping had lowered regional and riparian water tables below a level that willows could access with their roots (Kondolf 1996).

Planning restoration activities in the context of processes operating across diverse spatial scales is particularly critical because the great majority of drainage basins have multiple historical and contemporary human alterations and stressors (Frothingham et al. 2002). It may not be feasible to restore a formerly meandering river segment that is now braided, for example, without addressing continuing bank instability and high sediment inputs from former mining upstream (Hilmes and Wohl 1995). And it may not be feasible to restore a formerly braided river segment that now meanders through a densely vegetated river corridor without addressing the upstream flow regulation that prevents occurrence of peak flows capable of eroding banks and removing vegetation, as on the Platte River in Nebraska, USA (NRC 2004; Smith 2011) (Fig. 4.5).

Finally, thinking about restoration as applied to drainage basins or entire river networks rather than solely to limited segments of a river can help to prioritize the temporal and spatial scope of restoration activities. Examples come from the Upper Mississippi River and one of its principal tributaries, the Illinois River. Both rivers have been extensively and intensively altered to facilitate agriculture and settlement in valley bottoms (land drainage, floodplain forest clearance, levees, flood control) and navigation (locks and dams, dredging, channelization) (Wohl 2004, 2011, 2013). Agriculture, urbanization, and navigation continue in these river networks, but restoration activities focus on what are sometimes referred to as beads along these rivers.

Fig. 4.5 The Platte River east of Kearney, Nebraska, USA in 2004. The river here drains more than 170,000 km². Multiple shallow channels branch among densely vegetated small islands and stream banks. *Yellow arrow* indicates flow direction. (Photograph courtesy of Will Graf)

'Beads' was first used in connection with rivers by Stanford et al. (1996), who describe floodplains in bedrock canyons as resembling beads on a string when viewed in planform. Beads are the relatively wide, low gradient river segments with well-developed floodplains. Strings are the intervening steep, narrow segments with minimal floodplains. The pattern of longitudinal variations in valley and channel geometry is quite common along many rivers and has been used in the Upper Mississippi and Illinois River basins in the United States to target sites for restoration. Beads tend to be especially ecologically rich: these segments provide more abundant and diverse channel and floodplain habitat, have greater hyporheic exchange, and support greater biodiversity (Bellmore and Baxter 2014; Hauer et al. 2016). Beads are also agriculturally rich sites with intensive land use, but individual beads can be acquired and restored by notching levees and allowing riparian water tables to rise and native plant communities to regrow (Florsheim and Mount 2002). The spatially limited characteristics of beads makes it more feasible to acquire them for national wildlife refuges such as the Upper Mississippi River National Fish and Wildlife Refuge or private reserves such as The Nature Conservancy's Spunky Bottoms and Emiquon preserves on the Illinois River. Although the beads are of limited spatial extent, the cumulative effect of restoring multiple beads throughout a river network can substantially increase ecosystem services (Stanford et al. 1996; Galat et al. 1998).

Integrative planning that considers changes through time is equally important for at least three reasons. First, river networks can require centuries to recover from human alterations, even after the human activities have ceased. Sediment slugs in southeastern Australia, mill dams in the eastern United States, and lack of spawning gravels in the Oregon Coast Range, USA provide examples.

Clearing of native vegetation by settlers of European descent occurred throughout southeastern Australia between circa 1790 and 1850 (Brierley et al. 2005). As in other regions of the world, removal of upland vegetation caused increased sediment yields from uplands to river corridors. Removal of riparian vegetation and instream large wood reduced physical complexity and flow resistance within river corridors. In response, channels became wider, shallower, and straighter, with floodplain aggradation and development of inset benches where sediment accumulated along the margins of the active channel (Brooks and Brierley 1997; Fryirs and Brierley 2001; Brierley et al. 2005) (Fig. 4.6), as well as plugs of sand at tributary junctions that create shallow lakes (Robson and Mitchell 2010). River management and restoration in these river networks must account for continuing responses by the river corridor to historical land uses. In this example, sediment dynamics within river corridors continue to reflect land uses that occurred more than a century ago.

Mill dams provided power sources for numerous activities prior to the development of steam and electrical power. Working in the Mid-Atlantic Piedmont region of the United States, Walter and Merritts (2008) document the existence of thousands of mill dams built in close proximity to one another along many rivers, with the backwater from one dam touching the base of the dam upstream. Each mill pond collected sediment coming from uplands that were simultaneously being cleared for agriculture. As other power sources replaced water mills and the mill ponds completely filled with sediment, the dams were gradually breached and largely forgotten. Rivers cut down into the thick sequences of 'legacy sediment' that had accumulated along the river corridors and people forgot that the dams had ever been present. Gravel-bed rivers flowing between tall banks of finer sediment topped with riparian trees came to be viewed as the natural condition for the region until Walter and Merritts (2008) showed that many of the channels were historically swampy or marshy swales prior to construction of mill dams. Of particular importance, this work and subsequent studies (Schenk and Hupp 2009; Pizzuto and O'Neal 2009; Merritts et al. 2013) indicate that much of the fine sediment and nutrient loads causing problems in Chesapeake Bay and other estuaries originates primarily from erosion of these legacy sediments along river corridors rather than from erosion in upland crop fields. Consequently, river restoration designed to limit fine sediment and nutrient yields to downstream portions of the watershed has to focus on the presence and stability of river banks in addition to implementing best management practices in upland areas. In this example, channel form and sediment yields to downstream regions continue to reflect land use practices that have long since ceased.

Commercial timber harvest in the Oregon Coast Range of the western United States began during the early twentieth century and continued at high intensity until the 1990s (Gale et al. 2012). Clearcutting, particularly in this wet and tectonically active environment, substantially increases hillslope instability and water and sediment yields to river corridors. In the upper portions of river networks, debris flows

Fig. 4.6 (**a, b**) Downstream views of sand slugs and inset benches along the Macdonald River in the Hawkesbury River drainage basin of southeastern Australia in 2016. The active channel here is about 25 m wide and drains approximately 1800 km². The site is at 33.242°S, 150.940°E

triggered by clearcutting can scour river channels to bedrock, removing instream wood and cobble- to gravel-sized sediment that provides spawning habitat for salmonid fishes (May and Gresswell 2003). Instream wood can be particularly important in steep portions of a mountainous river network because the wood creates sufficient flow resistance and obstruction to trap coarse sediment and maintain an alluvial substrate in river segments that have bedrock beds in the absence of wood (Massong and Montgomery 2000). Even portions of the Coast Range where clearcutting ceased several decades ago continue to have relatively planar bedrock channels with minimal fish habitat because of the continuing high transport capacity for wood and sediment within the active channel. Effective river restoration must recognize these historical changes and account for limited retentive ability of the river segments.

Legacy effects refer generally to historical human alterations that continue to influence river ecosystems, even decades after the original human activity has ceased. General categories of legacy effects with respect to sediments (James 2013; Wohl 2015) include activities that reduce sedimentation within river corridors (e.g., levees that eliminate floodplain sedimentation), activities that enhance sedimentation (e.g., mill dams), and activities that contaminate river sediments with pollutants (Stanley and Doyle 2003). Other past alterations such as channel engineering are also sometimes described as leaving a legacy on river process and form.

A second reason to explicitly account for changes through time in planning river restoration is that some rivers undergo repeated alternations in process and form at timescales driven by external processes such as aperiodic precipitation extremes. An example comes from the semiarid steppe of the interior western United States. Rivers of the Great Plains alternate at time periods of several decades between braided, relatively unvegetated planform and meandering channels with a riparian forest. The change from meandering to braided occurs abruptly during a large rainfall-runoff flood that removes riparian forests and widens channels. The transition back to meandering occurs over a period of decades as native cottonwood (*Populus* spp.) trees that germinated immediately after the flood gradually stabilize the floodplain (Friedman and Lee 2002) (Fig. 4.7). Failure to recognize these repeated alterations in process and form within the river corridor can lead to inappropriate restoration targets (Kondolf et al. 2001).

A third reason to think about river restoration in the context of changes through time involves the future. Many, if not most, river basins will undergo continued alteration as a result of changing climate and human activities. Accounting for predicted changes in precipitation and proposed or likely future development may help to prioritize locations or forms of restoration within a drainage basin, as examined in more detail in the next section in the context of development space.

One of the most effective tools of integrative planning can be to create a conceptual model(s) of a river ecosystem and use that model to guide specific management actions. When combined with adaptive management that monitors river response to management, the conceptual model can be continually revised to reflect increasing understanding of the river ecosystem. Examples of conceptual models and adaptive

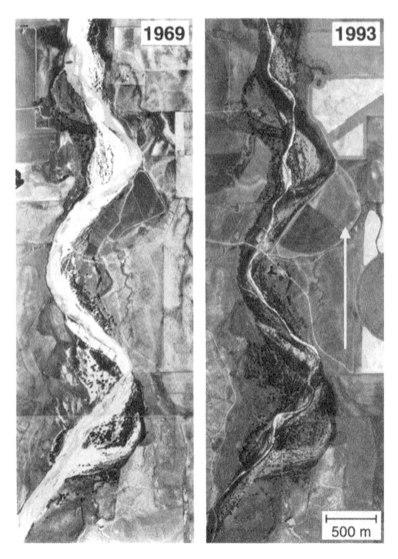

Fig. 4.7 Matched 1969 and 1993 historic air photos showing vegetation encroachment along Bijou Creek, Colorado after a large flood in 1965. The site is at 39.65°N, 104.217°W and the creek here drains 740 km². *Yellow arrow* indicates flow direction. (From Friedman and Lee 2002, Fig. 2).

management applied to river ecosystems come from the Apalachicola River basin in the southeastern United States (Richter et al. 2003), the Cooum River of India (Bunch and Dudycha 2004), rivers in Kruger National Park of South Africa (Rogers 2006), the Tisza River basin of southeastern Europe (Sendzimir et al. 2007), and the Platte River in the central United States (Smith 2011).

4.4 Response Curves and Development Space

King and Brown (2010) introduce the concept of development space in the context of relatively natural river networks. Approaching the issue of sustainable development, which can be defined as development that meets the needs of the present without compromising the ability of future generations to meet their own needs (WCED 1987), King and Brown propose an integrated basin flow assessment procedure to identify the optimum trade-off between costs and benefits of developing river resources through projects such as dams. The procedure involves evaluating multiple potential scenarios for future resource development, ranging from no development to complete development. The procedure also explicitly recognizes spatial variations that can result in a mosaic of different levels of river development and health, although the basin-wide effects of development in each portion of a river network are considered.

In order to evaluate different scenarios, King and Brown (2010) use response curves between physical drivers such as flow characteristics and response variables such as habitat area, abundance of particular aquatic species, or water temperature (Vanderpoorten and Durwael 1999) (Fig. 4.8). Response curves are quantitative where appropriate data exist and conceptual in situations where data are not available. Response curves in King and Brown (2010) use ratings of change from 0 for no change to 5 for severe change to describe predicted change in an ecosystem indicator to flow development (King et al. 2003). Response curves for numerous ecosystem characteristics are then combined into the decision support software Downstream Response to Imposed Flow Transformation (DRIFT) to evaluate differing scenarios of development (Brown and Joubert 2003; King et al. 2003) and to identify development space. Development space is the difference between current conditions in a river network and the furthest level of water-resource development found acceptable to stakeholders through consideration of the scenarios (King and Brown 2010; Fig. 4.9). Individual stakeholders will differ in what they consider acceptable development and consensus definitions of acceptable development will differ between river drainage basins. Nevertheless, the process provides a systematic method for considering the cumulative effects of multiple development projects, rather than discovering after all the projects have been built that the cumulative effects are unacceptable.

King et al. (2014) summarize an example of this type of analysis that was conducted for the Okavango River system in southern Africa. Using the information in 59 specialist reports on the 700,000 km² river basin, an environmental flows team divided the basin into 12 homogeneous biophysical and social units, 8 of which were identified as priorities for analysis because of proposed development. The team then identified 70 biophysical and 9 socioeconomic indicators responsive to flow changes and used 1100 response curves in DRIFT to analyze potential changes in the drainage basin under four scenarios of present-day conditions and low, medium, and high water-use development in future (King et al. 2014).

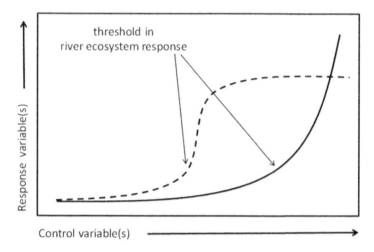

Fig. 4.8 Hypothetical response curves (after Wohl et al. 2015b, Fig. 4).

Fig. 4.9 Conceptual diagram illustrating development space (after King and Brown 2010, Fig. 4).

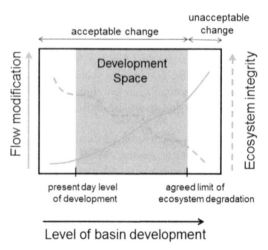

Response curves can also be employed in restoration designed to mitigate the effects of resource development that has already occurred. Prediction of physical and ecological changes in river ecosystems will never be as precise as constrained economic analyses because of complex interactions among the numerous processes within river ecosystems. However, enhanced predictive ability can result from development of response curves (Arthington et al. 2006; Merritt et al. 2010; Shafroth et al. 2010).

At least three challenges exist for using response curves to understand river ecosystem response to changes. The first challenge involves quantifying the relationship between control variables such as flow regime and response variables. The second challenge is that of understanding the limitations in applying the response curves. The third challenge is identifying all of the relevant variables at a site.

Quantifying the relationship between control and response variables can be difficult because the relationship is likely to be nonlinear and to include thresholds. Fish spawning in floodplain areas along the Ob River in Siberia, for example, requires at least 20 continuous days of overbank flows above a minimum discharge (Wohl 2011). Flows of smaller magnitude or shorter duration do not allow successful spawning. As another example, significant buffering of sediment and nitrate fluxes to a channel by riparian zones requires a minimum proportion of total channel length, minimum buffer width, and specific soil composition and plant community composition (Baker et al. 2006; Weller and Baker 2014). Riparian zones that do not meet these minima may still buffer sediment and nitrate fluxes to the channel, but the effect will be insignificant. Because thresholds in system function appear to be ubiquitous in many aspects of river process and form, identification of these thresholds using response curves can greatly improve the effectiveness of river restoration.

Limitations in applying response curves can arise from limited data used to develop the response curve, but a more important source of limitations is likely to result from the fact that response curves are bivariate, whereas river ecosystems are multivariate. For example, flow magnitude above some threshold might enhance fish spawning along the Ob River, but introduced species that feed on fish eggs and fry might nonetheless limit the successful reproduction of native fish. Consequently, an important limitation in using response curves is that the control variable being evaluated may not exert the most important limiting influence on the response variable.

Most response curves used thus far have focused on flow regime as the master variable of river ecosystems (Poff et al. 1997) and as the variable that is highly altered by human activities such as flow regulation. Restoring natural flow regime may not necessarily restore ecosystem function, however, if other important variables such as sediment regime (Wohl et al. 2015a), invasive organisms, or channel engineering limit river ecosystem response (Pasternack 2013). Identification of all relevant control variables can be difficult, but is critical to predicting and managing river ecosystem response to restoration.

Despite these challenges, response curves form a very useful tool for understanding river ecosystem behavior and are increasingly incorporated in planning environmental flows and other forms of river restoration.

4.5   Metrics of River Health

The European Union adopted a Water Framework Directive in 2000. Among other things, the directive promoted coordinated river basin management rather than individual countries acting independently to manage their particular portion of a river basin that crossed national boundaries. The Water Framework Directive also mandated good chemical (water quality) status and good ecological status for all surface waters by 2015. Metrics of water quality have existed for decades, but good ecological status is more nebulous. The directive defined ecological status in terms of the quality of the biological community and the hydrological and chemical characteristics of the river. Quality of the biological community was tied to the level of departure from the biological community that would be expected in conditions of minimal anthropogenic impact. Because rivers throughout Europe have been intensively manipulated—morphologically, hydrologically, chemically, and biologically—for centuries, true reference conditions are nearly impossible to find for many types of river ecosystems in Europe. Consequently, the Water Framework Directive initiated a surge of river research aimed at developing physical, biological, and integrative metrics of river health that could be used to assess departure from natural or reference conditions in diverse river ecosystems.

As noted in the first chapter, river health is an intuitively appealing concept that is easy to communicate at a general level to non-scientists (Karr 1999). Matters become more complicated when scientists attempt to quantify river health (Boulton 1999; Fairweather 1999; Harris and Silveira 1999). Much of the associated debate predates the EU Water Framework Directive and is in the biological literature because of the derivation of ideas of river health from ideas of ecosystem health (Norris and Thoms 1999).

Biologists commonly include physical and chemical characteristics of rivers when considering river health, as illustrated in the definition of river health as the degree to which a river's energy source, water quality, and flow regime, as well as the river's biota and habitats, match the natural conditions at all scales (Karr 1991; Harris and Silveira 1999). Many of the metrics of river health nonetheless focus on biological characteristics, partly because the metrics were developed by biologists and partly because the assumption is that the biota reflect or integrate physical and chemical characteristics of the river ecosystem. Examples include multimetric indices such as the Index of Biotic Integrity, which employs fish-community attributes of species richness, abundance, community structure, and the health of individual fish (Harris and Silveira 1999), and multivariate statistical methods used to discern pattern in taxonomic composition of macroinvertebrates or other organisms (Karr 1999). Biological metrics commonly focus on some aspect of biodiversity, which is typically defined in terms of number of species within a given ecosystem, but can be quantified in diverse ways (Gaston and Spicer 2004). Most of the metrics of river health described above focus on structural ecosystem components such as community composition (Estevez et al. 2017). Some scientists argue that river health is more effectively measured by integrating measures of structure with functional indicators such as nutrient retention or river metabolism (Bunn et al. 1999).

Cumulative metrics of river health can include measures of water quality (Bunn et al. 1999), habitat (Maddock 1999; Norris and Thoms 1999), or flow regime (Richter et al. 1996). Physical scientists, however, have been slower to develop metrics of physical river condition that can be used to assess departure from reference conditions, although the Water Framework Directive spurred research on this topic (Newson and Large 2006). Geomorphic conceptualizations of river health typically emphasize diversity of form and process through space and time (McDonald et al. 2004; Brierley and Fryirs 2005), including the ability to adjust form and process in response to changes in water and sediment inputs (Wohl 2012). Assessing the ability of a river corridor to geomorphically adjust to changing inputs is closer to assessing river functionality as proposed by Palmer and Febria (2012).

Regardless of the metrics used to assess river health, over-simplifying or standardizing the criteria for judging a river to be healthy must be avoided (Brierley and Fryirs 2009). Some rivers are naturally depauperate in species because of harsh physical or chemical conditions or a history of geographic isolation. Some rivers naturally receive frequent large inputs of sediment because of tectonic uplift, a semiarid climate, or erodible bedrock lithology. In these river corridors, the abundant sediment supply causes continual channel instability that can limit biodiversity. Some rivers episodically change their form and function in response to naturally occurring large floods. Other rivers may be fed predominantly by ground water springs that create a very stable and consistent flow regime. In both cases, biodiversity may be lower than in river corridors with an intermediate level of disturbance (Townsend et al. 1997). The key point is to recognize the levels of complexity, connectivity, and change naturally inherent in a particular river context.

4.6   Changing Perceptions of River Health and Naturalness

Not everyone views river corridors in the same way. As noted in the introductory chapter, river scientists from different disciplines typically have disparate emphases in conceptualizing rivers, although good progress has been made within the scholarly community in transcending disciplinary boundaries. River scientists, resource managers, and indigenous communities, as well as stakeholders who use the river for recreation versus those who rely on the river for water supply, may all have very different perceptions of the river corridor and ideas about how it should be managed (Eden et al. 2000). Many people, however, are moving beyond what might be termed mainstream attitudes toward river corridors that have developed over the past two to three centuries in high-income countries. These mainstream attitudes emphasize physical simplicity, uniformity, and human control of river corridors in the context of maximizing water supplies for consumptive uses and minimizing natural hazards associated with rivers.

Some of this shift in attitude is driven by the recognition that ecosystem services provided by rivers, which are critical to sustaining human societies, are declining sharply in most river basins around the world. Some of the shift in attitudes toward rivers is driven by a return within high-income societies to a recognition of the inherent right of non-human organisms and ecosystems to exist. New Zealand, for example, recognizes a river as a person in a legal context (Roy 2017). Some of the shift is driven by the realization that completely natural river corridors have largely ceased to exist, which can result in a longing for a more natural environment. Whatever the combination of forces driving a shift in management of river corridors, the shift is manifested in several recent trends. These trends include creating and sustaining physical complexity and connectivity within river corridors by deliberately reintroducing large wood (US Bureau of Reclamation 2016) in forested regions or beaver in the northern hemisphere (Pollock et al. 2015). Another trend in river management is mandating water quality standards that can maintain aquatic life as well as provide human drinking water (e.g., US Environmental Protection Agency, EU Water Framework Directory). A third trend involves removing dams or modifying dam operations (Bednarek 2001; Grant 2001; Stanley and Doyle 2003; O'Connor et al. 2015). Creating space for rivers to adjust to changing water and sediment inputs by setting back or notching levees and altering grade-control structures (Florsheim and Mount 2002; Moritsch 2017; Dutch Room for the River Programme, www.ruimtevoorderivier.nl) is a fourth trend in river management. Finally, becoming more conscious of the need to accept and embrace complexity, connectivity, and change within river corridors is a relatively recent shift in river management (Ward et al. 2001; Elosegi et al. 2010).

In Asia, Africa, and Latin America, however, enormous numbers of large dams are rapidly being built or are currently planned for construction. In the Mekong River basin, for example, seven large dams are under construction and an additional 133 are proposed (Kondolf et al. 2014). In the Amazon River basin, 151 new dams are proposed for the Andean portion of the western Amazon (Finer and Jenkins 2012). The budgets for large dams tend to be systematically biased below actual costs (Ansar et al. 2014). Governments and some international aid agencies view large dams favorably because the dams are believed to provide clean or green energy with minimal use of fossil fuels and because they are large projects that attract major funding and investors and can create a sense of national achievement and pride. In reality, construction of these dams is perpetuating the same devastation wrecked on river ecosystems in high-income countries during the twentieth century because the basic designs and operating regimes have varied little with respect to preserving riverine connectivity or complexity. High-income countries that have benefited economically from past dam construction can provide the research that might be used to modify traditional dam design, but such research is not being widely used. The rapid pace of traditional dam construction is especially disappointing given the numerous alternative means of generating energy, from micro-hydro power plants (Paish 2002) and within-channel turbines (Vermaak et al. 2014) to solar, wind, and biofuels. However, if rivers are viewed as simple channels to convey water downstream, rather than as ecosystems, then it does not matter much when or how the water is conveyed or whether it flows downstream at all.

Protecting river ecosystems remains a substantial challenge even in high-income countries with extensive legislation, standards, and programs designed to protect and restore rivers. An example comes from the increasing use of stream mitigation banking in the United States. Stream mitigation banking gives developers the option to offset construction impacts to streams by purchasing credits generated by for-profit companies that restore degraded streams (Lave et al. 2008). This is an unfortunate trend for several reasons. First, the requirements for establishing a stream mitigation credit market in a given region are not specified. Second, different definitions exist for a stream mitigation unit: some rely on quantity (e.g., linear length of stream), whereas others use minimally specified measures of quality (e.g., maintenance of cross-sectional form without substantial erosion or deposition over a specified time period). Third, existing stream mitigation projects have not required aquatic ecological assessment or monitoring, which can allow practitioners to employ only the most basic physical measures of a stream's form, rather than an integrative measure of stream function. Finally, if fully functional stream ecosystems are being degraded under the assumption that there is no net loss because of mitigation elsewhere, this is misleading. Even the most well designed and thorough restoration projects commonly do not fully restore stream ecosystems and local restoration projects do not compensate for basin-scale changes in factors such as land use (Bernhardt and Palmer 2011).

The challenge of protecting and restoring river ecosystems is enormous in a world where human population and energy and resource consumption are soaring, especially if that world is at least partly controlled by people who do not recognize the importance of maintaining functional river ecosystems. This challenge is common to any form of ecosystem protection and restoration or efforts to achieve ecological sustainability. Four examples of river basin management and restoration programs from Europe and North America illustrate efforts to address this challenge.

As noted previously, the EU Water Framework Directive mandates international, coordinated river basin management. The Danube River drainage basin includes the territories of 18 nations and has thus needed at least some level of international cooperation for many decades. As reviewed in Nachtnebel (2000), a basin-wide agreement governing navigation was developed after the Second World War, but other forms of cooperation prior to the 1990s were bilateral agreements between neighboring countries. The Environmental Programme for the Danube River Basin was conceived in 1991 as a means of addressing problems that included high nutrient loads; changes in flow and sediment transport regimes resulting from numerous dams; water pollution; and competition for water among consumptive users. The environmental action plan included priority actions related to water quality and river restoration (Botterweg and Rodda 1999; Bloesch and Sieber 2003), as well as studies in preparation for agreement on new regulations within the drainage basin. The first Danube River basin management plan was agreed on in 2009 (Sommerwerk et al. 2010). Progress thus far has included reducing nitrogen loads entering the Black Sea from the Danube basin (Balana et al. 2011) and generally improving water quality in the upper drainage basin (Wohl 2011). The management plan has also established a basin-wide monitoring network and restored connectivity and complexity along some floodplain reaches of the Danube and its major tributaries (Hohensinner et al. 2005).

A much smaller-scale example of river restoration that includes complexity and connectivity comes from the Mareiterbach, or Rio Ridanna, in the South Tyrol region of Italy. The Mareiterbach drains 212 km^2 of glaciated, mountainous terrain underlain by slate, which produces high sediment yields in this region of steep terrain and relatively wet climate (mean annual precipitation of ~800 mm) (Moritsch

2017). Land use in the valley bottom replaced riparian forests with agricultural lands and settlements, which led to more than 50% reduction in the spatial extent of the river corridor. Restoration designed to decrease flood risk and increase sustainable river habitat was undertaken during 2013–2015, with a focus on widening the active channel, decreasing the steepness of the banks along the incised channel, and removing grade-control structures (Fig. 4.10). Observed changes thus far include greater longitudinal connectivity for sediment transport, increased instream habitat diversity, and increased size and abundance of riparian vegetation patches (Moritsch 2017).

In North America, basin-scale restoration programs have included reconnecting channels with portions of the river corridor disconnected through channel engineering and/or flow regulation. Examples come from the Kissimmee River in Florida, USA and the Colorado River in the southwestern United States and northwestern Mexico.

The Kissimmee River is the headwaters of the Everglades ecosystem and the primary tributary to Lake Okeechobee. The river drains ~6100 km^2 of low-relief terrain with extensive wetlands (Warne et al. 2000). The highly sinuous Kissimmee was channelized and regulated with six water-control structures between 1962 and 1971. Nutrient loading to Lake Okeechobee increased, the river became anoxic, and habitat and wildlife populations declined. Public outcry led the Florida legislature to pass the Kissimmee River Restoration Act in 1976 with the goal of restoring ecological integrity as judged by energy source (organic matter inputs), water quality, habitat quality, hydrology, and biological interactions (Koebel 1995; Wohl 2004). Restoration included recreating a more natural flow regime by removing two of the water-control structures. Restoration also included enhancement of natural spatial heterogeneity via backfilling portions of the channelized canal and re-excavating buried portions of the original channel and removing barriers to connectivity created by artificially cutoff meanders (Fig. 4.11). Although the original hydrologic regime has not yet been restored and this limits full recovery of the river ecosystem (Toth et al. 1993), restoration activities completed thus far have produced the expected effects (Koebel and Bousquin 2014).

The Colorado River drains ~637,000 km^2 of the southwestern United States and northwestern Mexico, entering the Pacific Ocean in the Gulf of California. The river is intensively altered from the headwaters to the delta. Alterations include numerous dams for water storage and diversions for consumptive water uses within and beyond the drainage basin. The lower portion of the drainage basin is channelized. Agricultural return flows and salinity create water quality problems. Most of the flow regulation infrastructure was built during the second half of the twentieth century. By the end of the twentieth century it was clear that ecosystems throughout the river network were in serious decline. Restoration programs have focused on four portions of the river network: the Upper Colorado River basin upstream from the Grand Canyon; the Colorado River within the Grand Canyon; the Lower Colorado River basin downstream from the Grand Canyon; and the river's delta.

The Upper Colorado River endangered fish recovery program (http://www.coloradoriverrecovery.org/) emphasizes conservation and restoration of four endangered

Fig. 4.10 Aerial views of the Mareiterbach/Rio Ridanna, Italy before (2005, at *left*) and after (2010, at *right*) restoration. *Yellow arrow* indicates flow direction. The channel here drains about 200 km²; site is at 46.885°N, 11.379°E (Photographs courtesy of Francesco Comiti)

fish species that are endemic to the Colorado River basin. The program, which was initiated in 1988, includes numerous stakeholder groups who have worked together to create fish passages at migration barriers, as well as managing flows to provide spawning and rearing habitat in floodplain areas. The program also includes removal of nonnative fish species that compete with the endangered native species. Predation and competition by nonnative fish species, many of which were deliberately stocked in the river basin until the 1960s, now constitute the greatest threat to endangered fish recovery in the Upper Colorado (Highlights 2016).

Environmental management within the Grand Canyon portion of the Colorado River has received the most attention globally because it focuses on a famous US national park and a UNESCO World Heritage site. The US Bureau of Reclamation, which operates Glen Canyon Dam (completed in 1963), established an environmental studies program for the river in 1982 because of concern about downstream effects from the dam on the Colorado River ecosystem in Grand Canyon. This was followed by federal legislation designed to protect the river ecosystem in 1992. This legislation mandated that adaptive management be used to monitor and assess the effects of dam operation. An adaptive management program was established in 1997 and has thus far conducted multiple experimental high-flow releases from Glen Canyon Dam in 1996, 2004, 2008, 2012, 2013, and 2016. Each of these

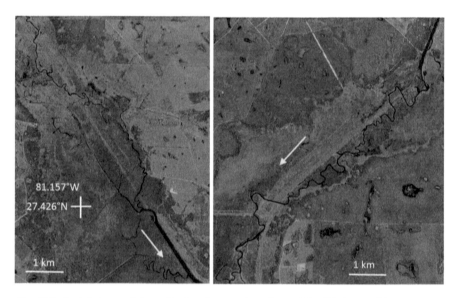

Fig. 4.11 Aerial views of channelized and restored portions of the Kissimmee River upstream from Lake Okeechobee, Florida in 2014. Portions of the river remaining channelized appear as broad, *straight dark lines*. Portions of the channelized river that have been backfilled appear as broad, straight, paler lines, with the restored sinuous channel to one side or crossing the backfilled canal. *Yellow arrows* indicate flow direction. (Photographs courtesy of Google Earth)

experimental flow releases received intense scrutiny with respect to physical and biological effects in the Colorado River ecosystem downstream from the dam (e.g., Melis 2011). Insights gained from each flow experiment have been used to refine the timing, magnitude, and duration of subsequent high-flow releases. The flow releases are primarily designed to redistribute continuing sediment inputs from tributaries entering the Colorado River downstream from Glen Canyon Dam in order to restore sand bars and associated backwater habitats (Fig. 4.12) for endangered native fish and other wildlife species. The planning, monitoring, and implementation processes associated with these experimental high flows have been used as models for experimental flow releases on other dammed rivers.

The Lower Colorado River Multi-Species Conservation Program (http://www.lcrmscp.gov/) focuses on river management and restoration below the Grand Canyon and within the United States (i.e., before the river passes into Mexico). The program includes a 50-year plan to conserve at least 26 species through implementation of a habitat conservation plan that returns physical complexity and connectivity to the river corridor, including managing flows to sustain these qualities (Bureau of Reclamation 2016).

As with many of the world's major river basins, the problems of the Colorado River drainage are perhaps most apparent at its delta, where the river largely ceased to flow to the ocean for a period of 16 years starting in 1998. Descriptions of the Colorado River delta during the early twentieth century indicate a lush subtropical

Fig. 4.12 2014 aerial views of backwaters along the Colorado River within Grand Canyon. The view at *left* shows an active backwater on the valley side of a pair of sand bars; the backwater remains connected to the main channel. At *right* is an inactive backwater that is now stabilized by riparian vegetation (likely nonnative, invasive tamarisk). (Photographs courtesy of Google Earth)

forest with extensive wetlands and abundant and diverse wildlife populations (Sykes 1937; Leopold 1949). Annual river discharges that could reach 7000 m³/s and transport substantial suspended sediment and organic matter supported habitat and wildlife (Cohen et al. 2001; Zamora et al. 2013). As water and sediment inputs ceased during filling of large upstream reservoirs during the 1930s–1960s, the delta became mostly a saline tidal flat. When the reservoir above Glen Canyon Dam finally reached capacity in 1981, flows began to again reach the delta during years of heavy precipitation in the drainage basin. These flows were small and of short duration, however, relative to historical flows to the delta. As of 2013, the river reached the ocean on average 12 days per year (Zamora et al. 2013). The United States has also discharged appreciable quantities of agricultural return flows into the southeastern portion of the delta since 1977, creating a large, brackish marsh known as Cienega de Santa Clara (Nelson et al. 2013) (Fig. 4.13). In an effort to further revitalize the delta, the US and Mexico agreed to an environmental flow release to the delta in spring 2014. The environmental effects of releasing 130 million m³ of water during this experiment were intensely monitored (Glenn et al. 2013) and included minor, localized erosion and deposition; germination of native vegetation; and ground water recharge (Mueller et al. 2016). The experimental delta flow was a small but symbolically important start to the process of revitalizing the Colorado River delta ecosystem.

The examples of the Danube, the Mareiterbach, the Kissimmee, and the Colorado each illustrate a shift in river management toward increasing physical complexity and connectivity of river corridors. Management of these rivers now focuses more

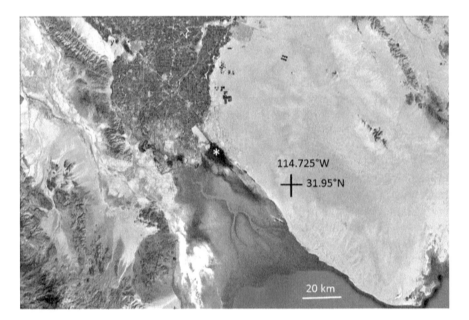

Fig. 4.13 2015 aerial view of the Colorado River delta on the Gulf of California. The *dark-green* patch marked with a *white asterisk* at the center of the photo is the wetland known as Cienega de Santa Clara, at 6000 ha the largest remaining wetland in the delta. The wetland is maintained by agricultural runoff from the Wellton-Mohawk Irrigation and Drainage District in Arizona and is monitored by a collaborative effort between the US and Mexico

on environmental flows and allowing the river corridor to adjust to fluctuations in water and sediment inputs by removing artificial constraints such as bank and bed stabilization. These types of projects can be expensive and require sustained, multi-year efforts for planning, implementation, monitoring, and adaptive management. The benefits from the activities may not be immediate or may be difficult to quantify, particularly in a monetary context. Increasing evidence suggests that benefits are present, however, and that these new management directions enhance river ecosystems (Pedersen et al. 2007; Opperman et al. 2010).

4.7   Where Next?

If societies treat river networks like canals, the result will be expensive, unruly canals that require continual maintenance and provide very little function beyond downstream conveyance of water. One of my favorite phrases in the scientific literature comes from a paper by my colleagues Kurt Fausch and Kevin Bestgen. Examining endangered native fish species in rivers of the U.S. Great Plains, where ground water pumping has lowered alluvial water tables and caused some streams to go dry, Fausch and Bestgen (1997) write "without water, there can be no fish." So simple, so obvious, so inescapably true, yet requiring expression in a part of the world where national legislation mandates saving endangered species but regional practice emphasizes unsustainable water consumption.

The Great Plains have a semiarid climate and the native vegetation is shortgrass steppe. This is not apparent when visiting communities such as the one I live in, which feature landscaping dominated by non-native plants that require supplemental irrigation to survive. Similarly, farms in the region predominantly grow irrigated crops that receive enormous amounts of water, while rivers that historically were perennial now nearly cease to flow during droughts. Regional population is growing at 6–9% per year and has been growing rapidly for decades, leading to increasing demand for water supplies that will not naturally increase and will almost certainly decrease as climate warms. As another of my favorite phrases in the scientific literature expresses the dilemma of steady increases in consumptive use through time, juxtaposed against the need to increase water allocations to river ecosystems, "How can this water be clawed back from other users?" (Arthington and Pusey 2003).

Water use in the Great Plains and throughout the western United States is governed by the doctrine of prior appropriation, which is sometimes paraphrased as 'use it or lose it.' Prior appropriation guarantees water rights for consumptive uses to the earliest (in historical time) legal claimants, but water users face the possibility of losing their full allocation if they do not consistently claim sufficient water each year. This policy tacitly endorses wasteful water consumption. A good starting point would be to effectively change the abbreviated description of prior appropriation to 'use it wisely or lose it' by mandating a consistent reduction, such as 10%, in water available to all legal claimants regardless of priority rights. Such a change can only occur if there is more widespread recognition that rivers are endangered ecosystems.

If we want our rivers to provide ecosystem services and to function as river ecosystems, rather than as drainage canals, we have to move beyond narrowly focused management that treats rivers as canals. This shift in approach is intellectually difficult because it requires more integrated and complex understanding than relying on a set of equations that describe flow energy and stable channel geometry. Managing rivers as ecosystems cannot rely on small-scale restoration projects undertaken without coordination across the entire watershed. Small projects conducted in isolation are less likely to result in watershed-scale benefits than systematic, coordinated targeting of portions of the river that are likely to restore the greatest ecosystem function. Managing rivers as ecosystems also cannot treat a river or the whole river network as if it exists independently of the watershed, including the subsurface. Critical controls on river process and form that exist outside of the management framework are likely to limit management success if these controls are not explicitly considered in planning. Managing rivers as ecosystems cannot operate under the usual project duration of a few months to a couple of years, with no monitoring or adaptive management that allows success or failure to inform subsequent projects. Short-term projects with no effective evaluation provide no basis for adapting management based on past successes or failures. Finally, managing rivers as ecosystems cannot succeed if the people living within the river basin do not value the river as an ecosystem that may not meet their ideals of natural beauty, but may nonetheless be fully functional.

If we want river ecosystem services *and sustainable water resources*, we must manage rivers as ecosystems and, for this, diverse sectors of the scientific community and the greater society must work together.

References

Abbe T, Brooks A (2011) Geomorphic, engineering and ecological considerations when using wood in river restoration. In: Simon A, Bennett SJ, Castro JM (eds) Stream restoration in dynamic fluvial systems: scientific approaches, analyses, and tools. American Geophysical Union Press, Washington, pp 419–451

Andrews ED, Nankervis JM (1995) Effective discharge and the design of channel maintenance flows for gravel-bed rivers. In: Costa JE et al (eds) Natural and anthropogenic influences in fluvial geomorphology. American Geophysical Union, Washington, pp 151–164

Ansar A, Flyvbjerg B, Budzier A, Lunn D (2014) Should we build more large dams? The actual costs of hydropower megaproject development. Energ Policy 69:43–56

Arthington AH, Pusey BJ (2003) Flow restoration and protection in Australian rivers. River Res Appl 19:377–395

Arthington AH, Bunn SE, Poff NL, Naiman RJ (2006) The challenge of providing environmental flow rules to sustain river ecosystems. Ecol Appl 16(4):1311–1318

Baker ME, Weller DE, Jordan TE (2006) Improved methods for quantifying potential nutrient interception by riparian buffers. Landsc Ecol 21:1327–1345

Balana BB, Vinten A, Slee B (2011) A review on cost-effectiveness of agri-environmental measures related to the EU WFD: key issues, methods, and applications. Ecol Econ 70:1021–1031

Bednarek AT (2001) Undamming rivers: a review of the ecological impacts of dam removal. Environ Manag 27:803–814

Beechie TJ, Sear DA, Olden JD, Pess GR, Buffington JM, Moir H, Roni P, Pollock MM (2010) Process-based principles for restoring river ecosystems. Bioscience 60:209–222

Bellmore JR, Baxter CV (2014) Effects of geomorphic process domains on river ecosystems: a comparison of floodplain and confined valley segments. River Res Appl 30:617–630

Bernhardt ES, Palmer MA (2011) River restoration—the fuzzy logic of repairing reaches to reverse watershed-scale degradation. Ecol Appl 21:1926–1931

Bernhardt ES, Palmer MA, Allan JD, and the National River Restoration Science Synthesis Working Group (2005) Restoration of U.S. rivers: a national synthesis. Science 308:636–637

Bernhardt ES et al (2007) Restoring rivers one reach at a time: results from a survey of U.S. river restoration practitioners. Restor Ecol 15:482–493

Bloesch J, Sieber U (2003) The morphological destruction and subsequent restoration programmes of large rivers in Europe. Arch Hydrobiol Suppl 147:363–385

Botterweg T, Rodda DW (1999) Danube River basin: progress with the environmental programme. Water Sci Technol 40:1–8

Boulton AJ (1999) An overview of river health assessment: philosophies, practice, problems and prognosis. Freshw Biol 41:469–479

Bovee KD, Milhous R (1978) Hydraulic simulation in instream flow studies: theory and techniques. Instream Flow Information Paper no. 5, FWS/OBS-78/33, U.S. Fish and Wildlife Service, Fort Collins, Colorado

Brierley GJ, Fryirs K (2000) River styles, a geomorphic approach to catchment characterization: implications for river rehabilitation in Bega catchment, New South Wales, Australia. Environ Manag 25:661–679

Brierley GJ, Fryirs KA (2005) Geomorphology and river management: applications of the river styles framework. Blackwell, Oxford, p 398

Brierley G, Fryirs K (2009) Don't fight the site: three geomorphic considerations in catchment-scale river rehabilitation planning. Environ Manag 43:1201–1218

Brierley GJ, Fryirs KA (2016) The use of evolutionary trajectories to guide 'moving targets' in the management of river futures. River Res Appl 32:823–835

Brierley GJ, Brooks AP, Fryirs K, Taylor MP (2005) Did humid-temperate rivers in the Old and New Worlds respond differently to clearance of riparian vegetation and removal of woody debris? Prog Phys Geogr 29:27–49

Brookes A (1990) Restoration and enhancement of engineered river channels: some European experiences. Regul Rivers Res Manag 5:45–56

Brooks AP, Brierley GJ (1997) Geomorphic responses of lower Bega River to catchment disturbance, 1851-1926. Geomorphology 18:291–304

Brooks AP, Howell T, Abbe TB, Arthington AH (2006) Confronting hysteresis: wood based river rehabilitation in highly altered riverine landscapes of south-eastern Australia. Geomorphology 79:395–422

Brown CA, Joubert A (2003) Using multicriteria analysis to develop environmental flow scenarios for rivers targeted for water resource development. Water SA 29:365–374

Bunch MJ, Dudycha DJ (2004) Linking conceptual and simulation models of the Cooum River: collaborative development of a GIS-based DSS for environmental management. Comput Environ Urban Syst 28:247–264

Bunn SE, Davies PM, Mosisch TD (1999) Ecosystem measures of river health and their response to riparian and catchment degradation. Freshw Biol 41:333–345

Bureau of Reclamation (US) (2016) Lower Colorado River Multi-Species Conservation Program: Balancing Resource Use and Conservation, Boulder City, NV. http://www.coloradoriverrecovery.org/

Campana D, Marchese E, Theule JI, Comiti F (2014) Channel degradation and restoration of an Alpine river and related morphological changes. Geomorphology 221:230–241

Chin A, Daniels MD, Urban MA, Piegay H, Gregory KJ, Bigler W, Butt AZ, Grable JL, Gregory SV, Lafrenz M, Laurencio LR, Wohl E (2008) Perceptions of wood in rivers and challenges for stream restoration in the United States. Environ Manag 41:893–903

Chin A et al (2009) Linking theory and practice for restoration of step-pool streams. Environ Manag 43:645–661

Chin A, Laurencio LR, Daniels MD, Wohl E, Urban MA, Boyer KL, Butt A, Piegay H, Gregory KJ (2014) The significance of perceptions and feedbacks for effectively managing wood in rivers. River Res Appl 30:98–111

Clifford NJ (2012) River restoration: widening perspectives. In: Church M, Biron P, Roy A (eds) Gravel-bed rivers: processes, tools, and environments. Wiley, Chichester, pp 150–159

Cohen MJ, Henges-Jeck C, Castillo-Moreno G (2001) A preliminary water balance for the Colorado River delta, 1992-1998. J Arid Environ 49:35–48

Collins BD, Montgomery DR, Haas AD (2002) Historical changes in the distribution and functions of large wood in Puget Lowland rivers. Can J Fish Aquat Sci 59:66–76

Comiti F, Mao L, Lenzi A, Siligardi M (2009) Artificial steps to stabilize mountain rivers: a post-project ecological assessment. River Res Appl 25:639–659

East AE et al (2015) Large-scale dam removal on the Elwha River, Washington, USA: river channel and floodplain geomorphic change. Geomorphology 228:765–786

Eden S, Bear C (2011) Reading the river through 'watercraft': environmental engagement through knowledge and practice in freshwater angling. Cult Geogr 18:297–314

Eden S, Tunstall SM (2006) Ecological versus social restoration? How urban river restoration challenges but also fails to challenge the science-policy nexus in the United Kingdom. Environ Plann C 24:661–680

Eden S, Tunstall SM, Tapsell SM (2000) Translating nature: river restoration as nature-culture. Environ Plann D 18:257–273

Elosegi A, Diez J, Mutz M (2010) Effects of hydromorphological integrity on biodiversity and functioning of river ecosystems. Hydrobiologia 657:199–215

Estevez E, Rodriguez-Castillo T, Alvarez-Cabria M, Penas FJ, Gonzalez-Ferreras AM, Lezcano M, Barquin J (2017) Analysis of structural and functional indicators for assessing the health state of mountain streams. Ecol Indic 72:553–564

Fairweather PG (1999) State of environment indicators of 'river health': exploring the metaphor. Freshw Biol 41:211–220

Fausch KD, Bestgen KR (1997) Ecology of fishes indigenous to the central and southwestern Great Plains. In: Knopf FL, Sampson FB (eds) Ecology and conservation of great plains vertebrates. Springer, New York, pp 131–166

Finer M, Jenkins CN (2012) Proliferation of hydroelectric dams in the Andean Amazon and implications for Andes-Amazon connectivity. PLoS One 7:e35126

Flessa KW, Glenn EP, Hinojosa-Huerta O, de le Parra-Renteria CA, Ramirez-Hernandez J, Schmidt JC, Zamora-Aroyo FA (2013) Flooding the Colorado River delta: a landscape-scale experiment. EOS Trans Am Geophys Union 94:485–486

Florsheim JL, Mount JF (2002) Restoration of floodplain topography by sand-splay complex formation in response to intentional levee breaches, lower Cosumnes River, California. Geomorphology 44:67–94

Friedman JM, Lee VJ (2002) Extreme floods, channel change, and riparian forests along ephemeral streams. Ecol Monogr 72:409–425

Frothingham KM, Rhoads BL, Herricks EE (2002) A multiscale conceptual framework for integrated ecogeomorphological research to support stream naturalization in the agricultural Midwest. Environ Manag 29:16–33

Fryirs K, Brierley GJ (2001) Variability in sediment delivery and storage along river courses in Bega catchment, NSW, Australia: implications for geomorphic river recovery. Geomorphology 38:237–265

Fryirs K, Brierley GJ (2009) Naturalness and place in river rehabilitation. Ecol Soc 14:20. http://www.ecologyandsociety.org/vol4/iss1/art20

Galat DL, Fredrickson LH, Humburg DD, Bataille KJ, Bodie JR et al (1998) Flooding to restore connectivity of regulated, large-river wetlands. Bioscience 48:721–733

Gale CB, Keegan CE, Berg EC, Daniels J, Christensen GA, Sorenson CB, Morgan TA, Polzin P (2012) Oregon's forest products industry and timber harvest, 2008: industry trends and impacts of the Great Recession through 2010. USDA Forest Service General Technical Report PNW-GTR-868, Portland, OR, p 55

Gao Y, Vogel RM, Kroll CN, Poff NL, Olden JD (2009) Development of representative indicators of hydrologic alteration. J Hydrol 374:136–147

Gaston KJ, Spicer JI (2004) Biodiversity: an introduction, 2nd edn. Blackwell, Hoboken, p 191

Glenn EP, Flessa KW, Pitt J (2013) Restoration potential of the aquatic ecosystems of the Colorado River Delta, Mexico: introduction to special issue on "Wetlands of the Colorado River Delta.". Ecol Eng 59:1–6

Grant (2001) Dam removal: panacea or Pandora for rivers? Hydrol Process 15:1531–1532

Gregory SV, Boyer KL, Gurnell AM (eds) (2003) Ecology and management of wood in world rivers. American Fisheries Society, Bethesda

Grigg NS (2016) Integrated water management: an interdisciplinary approach. Palgrave Macmillan, London

Gumiero B, Mant J, Hein T, Elso J, Boz B (2013) Linking the restoration of rivers and riparian zones/wetlands in Europe: sharing knowledge through case studies. Ecol Eng 56:36–50

Gurnell A, Tockner K, Edwards P, Petts G (2005) Effects of deposited wood on biocomplexity of river corridors. Front Ecol Environ 3:377–382

Harris JH, Silveira R (1999) Large-scale assessments of river health using an index of biotic integrity with low-diversity fish communities. Freshw Biol 41:235–252

Haslam SM (1994) River pollution: an ecological perspective. Wiley, Chichester

Hauer FR, Locke H, Dreitz VJ, Hebblewhite M, Lowe WH, Muhlfeld CC, Nelson CR, Proctor MF, Rood SB (2016) Gravel-bed river floodplains are the ecological nexus of glaciated mountain landscapes. Sci Adv 2:e1600026

Hester ET, Gooseff MN (2011) Hyporheic restoration in streams and rivers. In: Simon A, Bennett SJ, Castro JM (eds) Stream restoration in dynamic fluvial systems: scientific approaches, analyses, and tools. American Geophysical Union Press, Washington, pp 167–187

Hickford MJH, Schiel DR (2014) Experimental rehabilitation of degraded spawning habitat of a diadromous fish, *Galaxias maculatus* (Jenyns, 1842) in rural and urban streams. Restor Ecol 22:319–326

Highlights (2016) 2015-2016 highlights, upper Colorado endangered fish recovery program. US Department of Interior, Washington. http://www.coloradoriverrecovery.org/

Hilmes MM, Wohl EE (1995) Changes in channel morphology associated with placer mining. Phys Geogr 16:223–242

Hohensinner S, Jungwirth M, Muhar S, Habersack H (2005) Historical analyses: a foundation for developing and evaluating river-type specific restoration programs. Int J River Basin Manag 3:87–96

Hughes FMR, Rood SB (2003) Allocation of river flows for restoration of floodplain forest ecosystems: a review of approaches and their applicability in Europe. Environ Manag 32:12–33

Hughes FMR, Adams WM, Muller E, Nilsson C, Richards KS, Barsoum N, Decamps H, Foussadier R, Girel J, Guilloy H, Hayes A, Johansson M, Lamb L, Pautou G, Peiry JL, Perrow M, Vautier F, Winfield M (2001) The importance of different scale processes for the restoration of floodplain woodlands. Regul River Res Manag 17:325–345

James LA (2013) Legacy sediment: definitions and processes of episodically produced anthropogenic sediment. Anthropocene 2:16–26

Kail J, Hering D, Muhar S, Gerhard M, Preis S (2007) The use of large wood in stream restoration: experiences from 50 projects in Germany and Austria. J Appl Ecol 44:1145–1155

Karr JR (1991) Biological integrity: a long-neglected aspect of water resource management. Ecol Appl 1:66–84

Karr JR (1999) Defining and measuring river health. Freshw Biol 41:221–234

King J, Brown C (2010) Integrated basin flow assessments: concepts and method development in Africa and south-east Asia. Freshw Biol 55:127–146

King JM, Brown CA, Sabet H (2003) A scenario-based holistic approach to environmental flow assessments for regulated rivers. River Res Appl 19:619–640

King J, Beuster H, Brown C, Joubert A (2014) Pro-active management: the role of environmental flows in transboundary cooperative planning for the Okavango River system. Hydrol Sci J 59:786–800

Klösch M, Hornich R, Baumann N, Puchner G, Habersack H (2011) Mitigating channel incision via sediment input and self-initiated riverbank erosion at the Mur River, Austria. In: Simon A, Bennett SJ, Castro JM (eds) Stream restoration in dynamic fluvial systems: scientific approaches, analyses, and tools. American Geophysical Union Press, Washington, pp 319–336

Koebel JW (1995) An historical perspective on the Kissimmee River restoration project. Restor Ecol 3:149–159

Koebel JW, Bousquin SG (2014) The Kissimmee River restoration project and evaluation program, Florida, USA. Restor Ecol 22:345–352

Kondolf GM (1996) A cross section of stream channel restoration. J Soil Water Conserv 51:119–125

Kondolf GM (2006) River restoration and meanders. Ecol Soc 11(2):42. http://www.ecologyandsociety.org/vol11/isse/art42

Kondolf GM (2011) Setting goals in river restoration: when and where can the river 'heal itself'? In: Simon A, Bennett SJ, Castro JM (eds) Stream restoration in dynamic fluvial systems: scientific approaches, analyses, and tools. American Geophysical Union Press, Washington, pp 29–43

Kondolf GM, Smeltzer MM, Railsback SF (2001) Design and performance of a channel reconstruction project in a coastal California gravel-bed stream. Environ Manag 28:761–776

Kondolf GM, Boulton AJ, O'Daniel S, Poole GC, Rahel FJ, Stanley EH, Wohl E et al (2006) Process-based ecological river restoration: visualizing three-dimensional connectivity and dynamic vectors to recover lost linkages. Ecol Soc 11(2):5. http://www.ecologyandsociety.org/vol11/iss2/art5

Kondolf GM, Rubin ZK, Minear JT (2014) Dams on the Mekong: cumulative sediment starvation. Water Resour Res 50:5158–5169

Konrad CP, Black RW, Voss F, Neale CMU (2008) Integrating remotely acquired and field data to assess effects of setback levees on riparian and aquatic habitats in glacial-melt water rivers. River Res Appl 24:355–372

Konrad CP, Olden JD, Lytle DA, Melis TS, Schmidt JC et al (2011) Large-scale flow experiments for managing river systems. Bioscience 61:948–959

Kozak JP, Bennett MG, Piazza BB, Remo JWF (2016) Towards dynamic flow regime management for floodplain restoration in the Atchafalaya River basin, Louisiana. Environ Sci Policy 64:118–128

Kristensen EA, Kronvang B, Wiberg-Larsen P, Thodsen H, Nielsen C, Amor E, Friberg N, Pedersen ML, Baattrup-Pedersen A (2014) 10 years after the largest river restoration project in northern Europe: hydromorphological changes on multiple scales in River Skjern. Ecol Eng 66:141–149

Lave R, Robertson MM, Doyle MW (2008) Why you should pay attention to stream mitigation banking. Ecol Restor 26:287–289

Leopold A (1949) A sand county Almanac. Oxford University Press, New York

Lepori F, Palm D, Brannas E, Malmqvist B (2005) Does restoration of structural heterogeneity in streams enhance fish and macroinvertebrate diversity? Ecol Appl 15:2060–2071

Lester RE, Boulton AJ (2008) Rehabilitating agricultural streams in Australia with wood: a review. Environ Manag 42:310–326

Lorenz AW, Jahnig SC, Hering D (2009) Re-meandering German lowland streams: qualitative and quantitative effects of restoration measures on hydromorphology and macroinvertebrates. Environ Manag 44:745–754

Maddock I (1999) The importance of physical habitat assessment for evaluating river health. Freshw Biol 41:373–391

Major JJ et al (2012) Geomorphic response of the Sandy River, Oregon, to removal of Marmot Dam. U.S. Geological Survey Professional Paper 1792, p 64

Marks CO, Nislow KH, Magilligan FJ (2014) Quantifying flooding regime in floodplain forests to guide river restoration. Elementa Science Anthropocene 2:31. doi:10.12952/journal.elementa.000031

Massong TM, Montgomery DR (2000) Influence of sediment supply, lithology, and wood debris on the distribution of bedrock and alluvial channels. Geol Soc Am Bull 112:591–599

May CL, Gresswell RE (2003) Processes and rates of sediment and wood accumulation in headwater streams of the Oregon Coast Range, USA. Earth Surf Process Landf 28:409–424

McDonald A, Lane SN, Haycock NE, Chalk EA (2004) Rivers of dreams: on the gulf between theoretical and practical aspects of an upland river restoration. Trans Inst Brit Geogr 29:257–281

Melis TS (ed) (2011) Effects of three high-flow experiments on the Colorado River ecosystem downstream from Glen Canyon Dam, Arizona. U.S. Geological Survey Circular 1366, Reston, VA

Merritt DM, Scott ML, Poff NL, Auble GT, Lytle DA (2010) Theory, methods and tools for determining environmental flows for riparian vegetation: Riparian vegetation-flow response guilds. Freshw Biol 55(1):206–225

Merritts D, Walter R, Rahnis M, Cox S, Hartranft J, Scheid C, Potter N et al (2013) The rise and fall of Mid-Atlantic streams: millpond sedimentation, milldam breaching, channel incision, and stream bank erosion. In: DeGraff JV, Evans JE (eds) The challenges of dam removal and river restoration. Geological Society of America Reviews in Engineering Geology XXI, Boulder, pp 183–203

Moritsch S (2017) Influence on the riparian vegetation as a geomorphic quality unit at restored and degraded stretches of a highly dynamic river in South Tyrol. MS thesis, Free University of Bozen-Bolzano, Italy

Mueller ER, Grams PE, Schmidt JC, Hazel JE, Alexander JS, Kaplinski M (2014) The influence of controlled floods on fine sediment storage in debris fan-affected canyons of the Colorado River basin. Geomorphology 226:65–75

Mueller ER, Schmidt JC, Topping DJ, Shafroth PB, Rodriguez-Burgueno JE, Ramirez-Hernandez J, Grams PE (2016) Geomorphic change and sediment transport during a small artificial flood in a transformed post-dam delta: the Colorado River delta, United States and Mexico. Ecological Engineering

Muotka T, Laasonen P (2002) Ecosystem recovery in restored headwater streams: the role of enhanced leaf retention. J Appl Ecol 39:145–156

Mürle U, Ortlepp J, Zahner M (2003) Effects of experimental flooding on riverine morphology, structure and riparian vegetation: the River Spöl, Swiss National Park. Aquat Sci 65:191–198

Nachtnebel H-P (2000) The Danube River basin environmental programme: plans and actions for a basin wide approach. Water Policy 2:113–129

Nakamura K, Tockner K, Amano K (2006) River and wetland restoration: lessons from Japan. Bioscience 56:419–429

Nakamura F, Ishiyama N, Sueyoshi M, Negishi JN, Akasaka T (2014) The significance of meander restoration for the hydrogeomorphology and recovery of wetland organisms in the Kushiro River, a lowland river in Japan. Restor Ecol 22:544–554

Nelson SM, Fielding EJ, Zamora-Arroyo F, Flessa K (2013) Delta dynamics: effects of a major earthquake, tides, and river flows on Cienaga de Santa Clara and the Colorado River Delta, Mexico. Ecol Eng 59:144–156

Newson MD, Large ARG (2006) 'Natural' rivers, 'hydromorphological quality' and river restoration: a challenging new agenda for applied fluvial geomorphology. Earth Surf Process Landf 31:1606–1624

Norris RH, Thoms MC (1999) What is river health? Freshw Biol 41:197–209

NRC (National Research Council) (2004) Endangered and threatened species of the Platte River. National Research Council Board on Environmental Studies and Toxicology, Washington

O'Connor JE, Duda JJ, Grant GE (2015) 1000 dams and counting. Science 348:496–497

Opperman JJ, Luster R, McKenney BA, Roberts M, Meadows AW (2010) Ecologically functional floodplains: connectivity, flow regime, and scale. J Am Water Resour Assoc 46:211–226

Ortlepp J, Murle U (2003) Effects of experimental flooding on brown trout (*Salmo trutta fario* L.): the River Spöl, Swiss National Park. Aquat Sci 65:232–238

Paish O (2002) Small hydro power: technology and current status. Renew Sust Energ Rev 6:537–556

Palmer MA, Febria CM (2012) The heartbeat of ecosystems. Science 336:1393–1394

Palmer MA, Hondula KL (2014) Restoration as mitigation: analysis of stream mitigation for coal mining impacts in southern Appalachia. Environ Sci Technol 48:10552–10560

Palmer MA, Bernhardt ES, Allan JD, Lake PS, Alexander G, Brooks S et al (2005) Standards for ecologically successful river restoration. J Appl Ecol 42:208–217

Palmer MA, Allan JD, Meyer J, Bernhardt ES (2007) River restoration in the twenty-first century: data and experiential knowledge to inform future efforts. Restor Ecol 15:472–481

Palmer MA, Bernhardt ES, Schlesinger WH, Eshleman KN, Foufoula-Georgiou E, Hendryx MS, Lemly AD, Likens GE, Loucks OL, Power ME, White PS, Wilcock PR (2010a) Mountaintop mining consequences. Science 327:148–149

Palmer MA, Menninger HL, Bernhardt E (2010b) River restoration, habitat heterogeneity and biodiversity: a failure of theory or practice? Freshw Biol 55:205–222

Palmer MA, Filoso S, Fanelli RM (2014) From ecosystems to ecosystem services: stream restoration as ecological engineering. Ecol Eng 65:62–70

Pasternack GB (2013) A geomorphologist's guide to participating in river rehabilitation. In: Wohl E (ed) Treatise on fluvial geomorphology. Treatise on geomorphology, vol 9. Academic, San Diego, pp 843–860

Pedersen ML, Andersen JM, Nielsen K, Linnemann M (2007) Restoration of Skern River and its valley: project description and general ecological changes in the project area. Ecol Eng 30:131–144

Petkovska V, Urbanic G, Mikos M (2015) Variety of the guiding image of rivers—defined for ecologically relevant habitat features at the meeting of the alpine, Mediterranean, lowland and karst regions. Ecol Eng 81:373–386

Pettit NE, Naiman RJ (2006) Flood-deposited wood creates regeneration niches for riparian vegetation on a semi-arid South African river. J Veg Sci 17:615–624

Pfadenhauer J (2001) Some remarks on the socio-cultural background of restoration ecology. Restor Ecol 9:220–229

Pitlick J, Wilcock PR (2001) Relations between streamflow, sediment transport, and aquatic habitat in regulated rivers. In: Dorava JM et al (eds) Geomorphic processes and riverine habitat. American Geophysical Union Press, Washington, pp 185–198

Pizzuto J, O'Neal M (2009) Increased mid-twentieth century riverbank erosion rates related to the demise of mill dams, South River, Virginia. Geology 37:19–22

Poff NL, Zimmerman JKH (2010) Ecological responses to altered flow regimes: a literature review to inform the science and management of environmental flows. Freshw Biol 55:194–205

Poff NL, Allan JD, Bain MB, Karr JR, Prestegaard KL, Richter BD, Sparks RE, Stromberg JC (1997) The natural flow regime: a paradigm for river conservation and restoration. Bioscience 47:769–784

Pollock MM, Lewallen G, Woodruff K, Jordan CE, Castro JM (eds) (2015) The beaver restoration guidebook: working with beavers to restore streams, wetlands, and floodplains, v. 1.0. US Fish and Wildlife Service, Portland, p 189

Rathbun SL, Merritt DM, Wohl EE, Sanderson JS, Knight HAL (2009) Characterizing environmental flows for maintenance of river ecosystems: North Fork Cache la Poudre River, Colorado. In: James LA, Rathbun SL, Whittecar GR (eds) Management and Restoration of Fluvial Systems with Broad Historical Changes and Human Impacts. Geological Society of America Special Paper 451, Boulder, Colorado, pp 143–157

Rhoads BL, Herricks EE (1996) Naturalization of headwater streams in Illinois: challenges and possibilities. In: Brookes A, Shields FD (eds) River channel restoration: guiding principles for sustainable projects. Wiley, Hoboken, pp 331–367

Rhodes HM, Closs GP, Townsend CR (2007) Stream ecosystem health outcomes of providing information to farmers and adoption of best management practices. J Appl Ecol 44:1106–1115

Richter BD, Baumgartner J, Powell J, Braun D (1996) A method for assessing hydrologic alteration within ecosystems. Conserv Biol 10:1163–1174

Richter BD, Mathews R, Harrison DL, Wigington R (2003) Ecologically sustainable water management: managing river flows for ecological integrity. Ecol Appl 13:206–224

Robson BJ, Mitchell BD (2010) Metastability in a river subject to multiple disturbances may constrain restoration options. Mar Freshw Res 61:778–785

Rogers KH (2006) The real river management challenge: integrating scientists, stakeholders and service agencies. River Res Appl 22:269–280

Roley SS, Tank JL, Stephen ML, Johnson LT, Beaulieu JJ, Witter JD (2012) Floodplain restoration enhances denitrification and reach-scale nitrogen removal in an agricultural stream. Ecol Appl 22:281–297

Roy EA (2017) New Zealand river granted same legal rights as human being. The Guardian, 26 March 2017. https://www.theguardian.com/world/2017/mar/16/new-zealand-river-granted-same-legal-rights-as-human-being. Accessed 23 May 2017

Rubin DM, Nelson JM, Topping DJ (1998) Relation of inversely graded deposits to suspended-sediment grain-size evolution during the 1996 flood experiment in Grand Canyon. Geology 26:99–102

Sanderson JS, Rowan N, Wilding T, Bledsoe BP, Miller WJ, Poff NL (2012) Getting to scale with environmental flow assessment: the watershed flow evaluation tool. River Res Appl 28:1369–1377

Schenk ER, Hupp CR (2009) Legacy effects of colonial millponds on floodplain sedimentation, bank erosion, and channel morphology, Mid-Atlantic, USA. J Am Water Resour Assoc 45:597–606

Schmidt JC, Wilcock PR (2008) Metrics for assessing the downstream effects of dams. Water Resour Res 44:W04404. doi:10.1029/2006WR005092

Sear DA (1994) River restoration and geomorphology. Aquat Conserv Mar Freshwat Ecosyst 4:169–177

Sendzimir J, Magnuszewski P, Balogh P, Vari A (2007) Anticipatory modeling of biocomplexity in the Tisza River basin: first steps to establish a participatory adaptive framework. Environ Model Softw 22:599–609

Shafroth PB, Wilcox AC, Lytle DA, Hickey JT, Andersen DC, Beauchamp VB, Hautzinger A, McMullen LE, Warner A (2010) Ecosystem effects of environmental flows: modelling and experimental floods in a dryland river. Freshw Biol 55:68–85

Shields FJ, Knight SS, Lizotte R, Wren DG (2011) Connectivity and variability: metrics for riverine floodplain backwater rehabilitation. In: Simon A, Bennett SJ, Castro JM (eds) Stream restoration in dynamic fluvial systems: scientific approaches, analyses, and tools. American Geophysical Union Press, Washington, pp 233–246

Smith CB (2011) Adaptive management on the central Platte River—science, engineering and decision analysis to assist in the recovery of four species. J Environ Manag 92:1414–1419

Sommerwerk N, Bloesch J, Paunovic M, Baumgartner C, Venohr M, Schneider-Jacoby M, Hein T, Tockner K (2010) Managing the world's most international river: the Danube River basin. Mar Freshw Res 61:736–748

Souchon Y, Sabaton C, Deibel R, Reiser D, Kershner J, Gard M, Katopodis C, Leonard P, Poff NL, Miller WJ, Lamb BL (2008) Detecting biological responses to flow management: missed opportunities; future directions. River Res Appl 24:506–518

Stalnaker C, Lamb BL, Henriksen J, Bovee K, Bartholow J (1995) The instream flow incremental methodology: a primer for IFIM. National Biological Service, US Department of the Interior, Biological Report no. 29, Fort Collins, Colorado

Stanford JA, Ward JV, Liss WJ, Frissell CA, Williams RN, Lichatowich JA, Coutant CC (1996) A general protocol for the restoration of regulated rivers. Regul Rivers Res Manag 12:391–413

Stanley EH, Doyle MW (2003) Trading off: the ecological effects of dam removal. Front Ecol Environ 1:15–22

Sykes G (1937) The Colorado Delta. American Geographical Society Special Publication, New York

Tharme RE (2003) A global perspective on environmental flow assessment: emerging trends in the development and application of environmental flow methodologies for rivers. River Res Appl 19:397–441

Thompson DM (2013) The quest for the golden trout: environmental loss and America's iconic fish. University Press of New England, Hanover

Tockner K, Schiemer F, Baumgartner C, Kum G, Weigand E, Zweimuller I, Ward JV (1999) The Danube restoration project: Species diversity patterns across connectivity gradients in the floodplain system. Regul Rivers Res Manag 15:245–258

Toth LA, Obeysekera JTB, Perkins WA, Loftin MK (1993) Flow regulation and restoration of Florida's Kissimmee River. Regul Rivers Res Manag 8:155–166

Townsend CR, Scarsbrook MR, Dolédec S (1997) The intermediate disturbance hypothesis, refugia, and biodiversity in streams. Limnol Oceanogr 42:938–949

US Bureau of Reclamation (2016) National large wood manual: assessment, planning, design, and maintenance of large wood in fluvial ecosystems: restoring process, function and structure. Bureau of Reclamation and Army Corps of Engineers, Washington, p 628

Vanderpoorten A, Durwael L (1999) Trophic response curves of aquatic bryophytes in lowland calcareous streams. Bryologist 102:720–728

Vermaak HJ, Kusakana K, Koko SP (2014) Status of micro-hydrokinetic river technology in rural applications: a review of literature. Renew Sust Energ Rev 29:625–633

Violin CR, Cada P, Sudduth EB, Hassett BA, Penrose DL, Bernhardt ES (2011) Effects of urbanization and urban stream restoration on the physical and biological structure of stream ecosystems. Ecol Appl 21:1932–1949

Wade RJ, Rhoads BL, Rodriguez J, Daniels M, Wilson D, Herricks EE, Bombardelli F, Garcia M, Schwartz J (2002) Integrating science and technology to support stream naturalization near Chicago, Illinois. J Am Water Resour Assoc 38:931–944

Walter RC, Merritts DJ (2008) Natural streams and the legacy of water-powered mills. Science 319:299–304

Ward JV, Tockner K, Uehlinger U, Malard F (2001) Understanding natural patterns and processes in river corridors as the basis for effective river restoration. Regul Rivers Res Manag 17:311–323

Warne AG, Toth LA, White WA (2000) Drainage-basin-scale geomorphic analysis to determine reference conditions for ecologic restoration—Kissimmee River, Florida. Geol Soc Am Bull 112:884–899

WCED (World Commission on Environment and Development) (1987) Report of the World Commission on Environment and Development: Our Common Future. Annex to Document A/42/427—Development and International Co-operation: Environment. Oxford University Press, Oxford, UK

Weissmann HZ, Könitzer C, Bertiller A (2009) Strukturen der Fliessgewässer in der Schweiz. BAFU (Bundesamt für Umwelt), Bern, Switzerland. [in German]

Weller DE, Baker ME (2014) Cropland riparian buffers throughout Chesapeake Bay watershed: Spatial patterns and effects on nitrate loads delivered to streams. J Am Water Resour Assoc 50:696–712

Wiele SM, Wilcock PR, Grams PE (2007) Reach-averaged sediment routing model of a canyon river. Water Resour Res 43:W02425. doi:10.1029/2005WR004824

Wilcox AC, O'Connor JE, Major JJ (2014) Rapid reservoir erosion, hyperconcentrated flow, and downstream deposition triggered by breaching of 38 m tall Condit Dam, White Salmon River, Washington. J Geophys Res Earth Surf 119:1376–1394

Wohl E (2004) Disconnected rivers: linking rivers to landscapes. Yale University Press, New Haven

Wohl E (2011) A world of rivers: environmental change on ten of the world's great rivers. University of Chicago Press, Chicago

Wohl E (2012) Identifying and mitigating dam-induced declines in river health: three case studies from the western United States. Int J Sed Res 27:271–287

Wohl E (2013) Wide rivers crossed: the South Platte and the Illinois of the American Prairie. University Press of Colorado, Boulder

Wohl E (2014) A legacy of absence: wood removal in U.S. rivers. Prog Phys Geogr 38:637–663

Wohl E (2015) Legacy effects on sediments in river corridors. Earth Sci Rev 147:30–53

Wohl E (2017) Bridging the gaps: an overview of wood across time and space in diverse rivers. Geomorphology 279:3–26

Wohl E, Angermeier PL, Bledsoe B, Kondolf GM, MacDonnell L, Merritt DM, Palmer MA, Poff NL, Tarboton D (2005) River restoration. Water Resour Res 41:W10301. doi:10.1029/2005WR003985

Wohl E, Bledsoe BP, Jacobson RB, Poff NL, Rathburn SL, Walters DM, Wilcox AC (2015a) The natural sediment regime in rivers: broadening the foundation for ecosystem management. Bioscience 65:358–371

Wohl E, Lane SN, Wilcox AC (2015b) The science and practice of river restoration. Water Resour Res 51:5974–5997

Yaning C, Qiang W, Weihong L, Xiao R, Yapeng C, Lihua Z (2006) Rational groundwater table indicated by the eco-physiological parameters of the vegetation: a case study of ecological restoration in the lower reaches of the Tarim River. Chin Sci Bull 51:8–15

Zamora HA, Nelson SM, Flessa KW, Nomura R (2013) Post-dam sediment dynamics and processes in the Colorado River estuary: implications for habitat restoration. Ecol Eng 59:134–143

Zhang L, Mitsch WJ (2007) Sediment chemistry and nutrient influx in a hydrologically restored bottomland hardwood forest in midwestern USA. River Res Appl 23:1026–1037

Glossary

Active channel The portion of a river corridor that contains water within a channel defined by the presence of stream bed and banks, which are typically delineated by erosional and depositional forms created by river processes; outside of drylands, the active channel typically does not support mature woody vegetation

Allochthonous energy sources Those derived from outside the channel, such as organic matter falling into the channel from the riparian forest or dissolved material carried into the channel by surface or subsurface runoff

Alternate stable states A conceptual model that describes ecological systems that can exist in multiple, distinct, and self-reinforcing states in equilibrium under equivalent environmental conditions

Anabranching A channel planform in which multiple, subparallel channels branch and rejoin downstream; individual channels are separated by vegetated floodplain or islands

Autochthonous energy sources Those derived from within the channel, such as photosynthesis by aquatic plants

Avulsion The rapid abandonment of a river channel and formation of a new channel, which is most likely to occur during a flood

Balanced sediment regime A balance between sediment supply and transport capacity that maintains specific physical processes and forms within the river corridor

Base level The lowest point in the landscape to which upstream portions of a river will erode; this can be sea level or an intermediate point within the river network, such as water level in a lake or the elevation of a main stem river that forms base level for a tributary channel

Bead (river) A river segment with a wider floodplain that has greater abundance and diversity of habitat and biota

Biota The animal and plant life of a river ecosystem

Biotic community Here used synonymously with biota

© The Author(s) 2018
E. Wohl, *Sustaining River Ecosystems and Water Resources*, SpringerBriefs in
Environmental Science, DOI 10.1007/978-3-319-65124-8

Braided A channel planform in which individual channels branch and rejoin downstream around bars and islands that commonly have relatively little woody riparian vegetation

Catchment See drainage basin

Channel maintenance flows Components of a river flow regime necessary to maintain specific channel characteristics, such as cross-sectional area for flood conveyance

Channel migration zone The width of the valley bottom across which channels can migrate and have migrated under the contemporary flow regime

Characteristic form time Describes the persistence of a feature within a river relative to the recurrence interval of the flow that creates the feature; a transient form is created by a disturbance, but then modified to pre-disturbance conditions before the recurrence of a disturbance of similar magnitude; a persistent form is present longer than the recurrence interval of the flood that created the form

Complexity Spatial heterogeneity, or the degree to which a river corridor deviates from a straight, uniform canal

Complex response A nonlinear geomorphic response to a single external change such as a large flood, change in land cover, or drop in base level

Connectivity The transfer of materials (e.g., water, sediment), energy, and organisms between components of a system; commonly described for rivers with respect to longitudinal, lateral, and vertical dimensions

Context The environmental setting of the river corridor; the climate and precipitation regime, geology, topography, biome, position in the river network, land use, and land use history

Development space In the context of water-resource development, development space is the difference between current conditions in a river network and the furthest level of water-resource development found acceptable to stakeholders through consideration of diverse scenarios of water development

Direct alterations (of rivers) Human activities undertaken within the river network and which affect river corridor configuration and connectivity, as well as water and sediment regimes

Disturbance Any relatively discrete event in time that disrupts ecosystem, community, or population structure and changes resources, substrate availability, or the physical environment

Disturbance regime The spatial pattern and statistical distribution of disturbances with respect to magnitude, frequency, and duration of associated changes in the physical environment

Downstream hydraulic geometry Conceptual model in which discharge is the dominant control on channel geometry and channel parameters such as width, depth, and velocity change progressively downstream as discharge increases

Drainage basin For any designated point, all of the river network and adjacent landscape that drains to that point; used synonymously with catchment and watershed

Ecological integrity The ability of the river corridor to support and maintain a community of organisms with species composition, diversity, and functional organization similar to those within natural habitats in the same region

Ecoregion A relatively large area of land or water that contains a geographically distinct assemblage of natural communities

Ecosystem A community of living organisms linked together and to the adjacent environment through fluxes of nutrients and energy

Ecosystem services Include the categories of provisioning (products obtained from ecosystems), regulating (benefits obtained from the regulation of ecosystem processes), supporting (services necessary for the production of all other ecosystem services), and cultural (non-material benefits that people derive from ecosystems)

Environmental flow Can refer to a specific flow release, such as an experimental flood release from a dam, or to an annual hydrograph that specifies magnitude, frequency, timing, duration, and rate of change in flow

Equilibrium A change to inputs or controlling parameters will result in a proportional change in the river corridor

Feedback Describes adjustments among components of a river ecosystem; positive or self-enhancing feedback occurs when an initial change creates a cascade of subsequent changes that amplify the initial change; negative or self-arresting feedback occurs when an initial change is dampened by the response of the river corridor

Floodplain That portion of the valley bottom inundated by water overflowing the channel banks during peak flows that occur every year or every few years

Flood-pulse model Describes the ecological influence of the seasonal flood pulse on large floodplain rivers such as the Amazon; water, sediment, nutrients, and organisms move from the channel onto the floodplain during peak flood flow and then return to the main channel and secondary channels during the receding limb and base flow

Flow-pulse model Flow pulses are fluctuations in discharge that change the extent of flow and standing water within a braided or anabranching channel segment; model emphasizes the ecological importance of these fluctuations

Hierarchical patch dynamics Characterizes river corridors as consisting of relatively homogeneous patches from the scale of microhabitat up to channel reaches, with distinct changes in process and form between patches

Hot moment Short period with disproportionately high biogeochemical reaction rates relative to longer intervening periods

Hot spot Patch with a disproportionately high biogeochemical reaction rate relative to the surrounding matrix

Hyporheic zone The area below the channel and floodplain in which flow paths originate from and terminate in the channel

Hypoxia Occurs when dissolved oxygen levels are ≤ 2 mg per liter, commonly as a result of excess nutrients that lead to high primary production and then decay of organic matter

Indirect alterations of rivers Human activities within the drainage basin but outside of the river corridor, such as changes in land cover, that affect inputs to the river corridor

Instream flow A specified minimum discharge within the active channel

Introduced species Non-native plants and animals accidentally or deliberately introduced to a river corridor; invasive introduced species are those that dramatically increase in population densities and geographic range after introduction

Lag time The time between a change in a control variable(s) and the response of some component of the river ecosystem

Legacy effect Historical human alterations that continue to influence river ecosystems

Legacy sediment Sediment that accumulates within a river corridor as a result of historical human activities, such as sediment filling the backwater of an abandoned dam

Natural flow regime The characteristics of the hydrograph present prior to intensive human alteration of a watershed; commonly described in terms of magnitude, frequency, duration, timing, and rate of change of flow

Natural (historical) range of variability The temporal variations in specified river parameters under natural conditions; for example, the range of active channel width through time within a particular river segment, assuming that channel width will vary in response to variations in water and sediment inputs

Natural sediment regime The characteristics of sediment inputs, transport, and storage present prior to intensive human alteration of a watershed

Nutrient spiraling Nutrients such as nitrogen and phosphorus are displaced downstream as they complete a cycle through the generalized compartments of water, particulates, and consumers; spiraling length refers to the downstream distance required for one complete cycle or, for organic carbon, as the distance between its entry into the river corridor and its oxidation

Particulate organic matter Organic matter carried in rivers is distinguished by size as fine particulate organic matter (0.45 μm–1 mm) and coarse particulate organic matter (>1 mm)

Physical integrity A set of active river processes and landforms such that the river corridor adjusts to changes in water and sediment inputs within limits of change defined by societal values

Process domains Spatially discrete portions of a landscape or river network that are characterized by distinct suites of geomorphic processes

Reference conditions Natural or reference conditions typically refers to the characteristics of the river ecosystem present prior to intensive human manipulation of the environment within the watershed and the river corridor

Resilience The persistence of the river ecosystem and its ability to absorb external changes and maintain the same relationships between biological populations or physical parameters

Resistance The ability of a river ecosystem to resist changes in form and process caused by disturbance

Response curve Graphical illustration of the relationship between a control variable, such as discharge, and a response variable, such as fish biomass

River continuum concept Progressive downstream changes in the relative importance of primary production versus respiration and in the structure and function of aquatic communities

River corridor Includes the main channel and secondary channels, where these are present; the floodplain; and the hyporheic zone underlying the channel and floodplain

River health Can be defined as the degree to which river corridor energy sources, water quality, flow regime, habitat, and biota match the natural conditions

River integrity The ability of the river to adjust to changing water and sediment inputs (without constraints imposed by human manipulation such as dams or levees) and through these adjustments to maintain the habitat, disturbance regime, and connectivity necessary to sustain native biotic communities

River metamorphosis An abrupt, sustained change in river form

River styles A geomorphic classification applied at the reach scale that is similar to process domains in recognizing spatially discrete reaches that are internally homogeneous with respect to topography, valley geometry, channel planform, and substrate, but differ from other reaches

Secondary channel Contains flowing water only during higher discharges

Sensitivity The ability of a river corridor to return to its pre-disturbance configuration following disturbance

Serial discontinuity model Dams and reservoirs reset the longitudinal trends described in the river continuum concept

Sustainability In an ecological context, typically refers to the ability of ecosystems to remain diverse and productive; in a physical context, may refer to the ability of a system to continue functioning or providing natural resources

Terrace (river) A former channel or floodplain surface now elevated above contemporary flood levels as a result of continued deposition on the floodplain or incision by the active channel

Threshold Separates distinct forms or modes of operation of a system; external thresholds refer to change forced by an external factor such as a flood; internal or intrinsic thresholds describe abrupt changes in a river corridor in the absence of external change

Watershed See drainage basin

Index

© The Author(s) 2018
E. Wohl, *Sustaining River Ecosystems and Water Resources*, SpringerBriefs in
Environmental Science, DOI 10.1007/978-3-319-65124-8

CPSIA information can be obtained
at www.ICGtesting.com
Printed in the USA
BVOW07s0917120917
494630BV00013B/25/P

9 783319 651231